A GUIDE TO SQL

Eighth Edition

A GUIDE TO SQL

Eighth Edition

Philip J. Pratt
Grand Valley State University

Mary Z. Last
University of Mary Hardin-Baylor

Australia • Brazil • Japan • Korea • Mexico • Singapore • Spain • United Kingdom • United States

A Guide to SQL, Eighth Edition
Philip J. Pratt, Mary Z. Last

Vice President, Publisher: Jack Calhoun

Editor-in-Chief: Alex von Rosenberg

Senior Acquisitions Editor: Charles McCormick, Jr.

Product Manager: Kate Hennessy

Development Editor: Jessica Evans

Editorial Assistant: Bryn Lathrop

Marketing Director: Brian Joyner

Marketing Manager: Bryant Chrzan

Marketing Communications Manager: Libby Shipp

Marketing Coordinator: Suellen Ruttkay

Content Project Manager: Matt Hutchinson

Art Director: Stacy Jenkins Shirley, Marissa Falco

Cover Designer: Joseph Sherman

Cover Image: Getty Images/Taxi/Chris Bell

Manufacturing Coordinator: Denise Powers

ISBN-13: 978-0-324-59768-4
ISBN-10: 0-324-59768-1

Course Technology
25 Thomson Place
Boston, MA, 02210
USA

Cengage Learning is a leading provider of customized learning solutions with office locations around the globe, including Singapore, the United Kingdom, Australia, Mexico, Brazil and Japan. Locate your office at: **international.cengage.com/region**

Cengage Learning products are represented in Canada by Nelson Education, Ltd.

For your lifelong learning solutions, visit **www.cengage.com**

Purchase any of our products at your local college store or at our preferred online store **www.ichapters.com**

Printed in the United States
2 3 4 5 6 7 12 11 10

TABLE OF CONTENTS

PREFACE

Structured Query Language (or SQL, which is pronounced “se-quel,” or “ess-cue-ell”) is a popular computer language that is used by diverse groups such as home computer users, owners of small businesses, end users in large organizations, and programmers. Although this text uses the SQL implementation in the Oracle Database 10*g* Express Edition as a vehicle for teaching SQL, its chapter material, examples, and exercises can be completed using any SQL implementation.

A Guide to SQL, Eighth Edition is written for a wide range of teaching levels, from students taking introductory computer science classes to those students in advanced information systems classes. This text can be used for a standalone course on SQL or in conjunction with a database concepts text where students are required to learn SQL.

The chapters in this text should be covered in order. Students should complete the end-of-chapter exercises and the examples within the chapters for maximum learning. Because the content of Chapter 8 assumes that the reader has had instruction or experience in at least one programming language, the instructor should determine whether students will understand its concepts. Students without a programming background will have difficulty understanding the topic of embedded SQL. Instructors can easily omit Chapter 8 from the text in situations where students are not comfortable with programming examples.

The Eighth Edition builds on the success of previous editions by presenting basic SQL commands in the context of a business that uses SQL to manage orders, parts, customers, and sales reps. Like in previous editions, this edition uses Oracle as the vehicle to present SQL commands. Like the last edition, this edition addresses SQL in Access™ by showing the Access versions of the same commands when they differ from the Oracle versions. This new edition also shows SQL Server commands when they differ from the Oracle versions. Differences for Access and SQL Server users are featured in “User” notes, which make it easy for students to identify differences for the SQL implementation they are using. Students can download the Oracle 10*g* Database Express Edition from the Oracle Web site for free and use it to complete this text without having to purchase or install the full Oracle program.

The Eighth Edition includes an entire chapter on database design, showing students how to create an appropriate design that satisfies a given set of requirements, and includes coverage of the important topics of stored procedures and triggers. The text also contains updated exercises for the Premiere Products, Henry Books, and Alexamara Marina Group cases.

DISTINGUISHING FEATURES

Use of Examples

Each chapter contains multiple examples that use SQL to solve a problem. Following each example, students will read about the commands that are used to solve the stated problem, and then they will see the SQL commands used to arrive at the solution. For most students, learning through examples is the most effective way to master material. For this reason, instructors should encourage students to read the chapters at the computer and input the commands shown in the figures.

Case Studies

A running case study—Premiere Products—is presented in all of the examples within the chapters and in the first set of exercises at the end of each chapter. Although the database is small in order to be manageable, the examples and exercises for the Premiere Products database simulate what a real business can accomplish using SQL commands. Using the same case study as examples within the chapter and in the end-of-chapter exercises ensures a high level of continuity to reinforce learning.

A second case study—the Henry Books database—is used in a second set of exercises at the end of each chapter. A third case study—the Alexamara Marina Group database—is used in a third set of exercises at the end of each chapter. The second and third case studies give students a chance to venture out "on their own" without the direct guidance of examples from the text.

Question and Answer Sections

A special type of exercise, called a Q&A, is used throughout the book. These exercises force students to consider special issues and understand important questions before continuing with their study. The answer to each Q&A appears after the question. Students are encouraged to formulate their own answers before reading the ones provided in the text to ensure that they understand new material before proceeding.

"User" Notes for Access™ and SQL Server Users

When an SQL command has a different use or format in Access or SQL Server, it appears in a User note. When you encounter a User note for the SQL implementation you are using, be sure to read its contents. You might also review the User notes for other SQL implementations so you are aware of the differences that occur from one implementation of SQL to another.

Review Material

A Summary and Key Terms list appear at the end of each chapter, followed by Review Questions that test students' recall of the important points in the chapter and occasionally test their ability to apply what they have learned. The answers to the odd-numbered Review Questions are provided in Appendix C. Each chapter also contains exercises related to the Premiere Products, Henry Books, and Alexamara Marina Group databases.

Appendices

Three appendices appear at the end of this text. Appendix A is an SQL reference that describes the purpose and syntax for the major SQL commands featured in the text. Students can use Appendix A to identify how and when to use important commands quickly. The SQL reference appendix contains references to specific pages in the text where the command is discussed to make it easy for students to find additional information when they need to refer back to the section in the book where the topic is covered.

Appendix B includes a "How Do I" reference, which lets students cross-reference the appropriate section in Appendix A by searching for the answer to a question. Appendix C includes answers to the odd-numbered Review Questions.

Relationship to Concepts of Database Management, Sixth Edition

For database courses featuring SQL, this SQL text can be bundled with *Concepts of Database Management, Sixth Edition* by Pratt and Adamski (Course Technology). The data and pedagogy between the two texts is consistent, and the instructor's manuals for both books include feedback and suggestions for using the texts together.

Instructor Support

The Eighth Edition includes a package of proven supplements for instructors and students. The Instructor's Resources offer a detailed electronic Instructor's Manual, figure files, Microsoft® PowerPoint® presentations, and the ExamView® Test Bank. The Instructor's Manual includes suggestions and strategies for using this text, as well as answers to Review Questions and solutions to the end-of-chapter exercises. Figure files allow instructors to create their own presentations using figures appearing in the text. Instructors can also take advantage of lecture presentations provided on PowerPoint slides; these presentations follow each chapter's coverage precisely, include chapter figures, and can be customized. ExamView is a powerful objective-based test generator that enables instructors to create paper, LAN, or Web-based tests from test banks designed specifically for this Course Technology text. Users can utilize the ultra-efficient QuickTest Wizard to create tests in less than five minutes by taking advantage of Course Technology's question banks, or can customize their own exams from scratch.

The Instructor's Resources include copies of the databases for the Premiere Products, Henry Books, and Alexamara Marina Group cases in Microsoft Access 2007 and 2003 formats and script files to create the tables and data in these databases in Oracle and SQL Server. These files are provided so instructors have the choice of assigning exercises in which students create the databases used in this text and load them with data, or they can provide the starting Access databases or Oracle or SQL Server script files to students to automate and simplify these tasks.

ORGANIZATION OF THE TEXT

The text contains eight chapters and three appendices, which are described in the following sections.

Chapter 1: Introduction to Premiere Products, Henry Books, and Alexamara Marina Group

Chapter 1 introduces the three database cases that are used throughout the text: Premiere Products, Henry Books, and Alexamara Marina Group. Many Q&A exercises are provided throughout the chapter to ensure that students understand how to manipulate the database on paper before they begin working in SQL.

Chapter 2: Database Design Fundamentals

Chapter 2 covers important concepts and terminology associated with relational databases, functional dependence, and primary keys, followed by a method for designing a database to satisfy a given set of requirements. It also illustrates the normalization process for finding and correcting a variety of potential problems in database designs. Finally, it shows how to represent database designs graphically using entity-relationship diagrams.

Chapter 3: Creating Tables

In Chapter 3, students begin using a DBMS by creating and running SQL commands to create tables, use data types, and add rows to tables. Chapter 3 also discusses the role of and use of nulls.

Chapter 4: Single-Table Queries

Chapter 4 is the first of two chapters on using SQL commands to query a database. The queries in Chapter 4 all involve single tables. Included in this chapter are discussions of simple and compound conditions; computed columns; the SQL BETWEEN, LIKE, and IN operators; using SQL aggregate functions; nesting queries; grouping data; and retrieving columns with null values.

Chapter 5: Multiple-Table Queries

Chapter 5 completes the discussion of querying a database by demonstrating queries that join more than one table. Included in this chapter are discussions of the SQL IN and EXISTS operators, nested subqueries, using aliases, joining a table to itself, SQL set operations, and the use of the ALL and ANY operators. The chapter also includes coverage of various types of joins.

Chapter 6: Updating Data

In Chapter 6, students learn how to use the SQL COMMIT, ROLLBACK, UPDATE, INSERT, and DELETE commands to update table data. Students also learn how to create a new table from an existing table and how to change the structure of a table. The chapter also includes coverage of transactions, including both their purpose and implementation.

Chapter 7: Database Administration

Chapter 7 covers the database administration features of SQL, including the use of views; granting and revoking database privileges to users; creating, dropping, and using an index; using and obtaining information from the system catalog; and using integrity constraints to control data entry.

Chapter 8: SQL Functions and Procedures

Chapter 8 begins with a discussion of some important SQL functions that act on single rows. Students will also learn how to use PL/SQL and T-SQL to cover the process of embedding SQL commands in another language. Included in this chapter are discussions of using embedded SQL to insert new rows and change and delete existing rows. Also included is a discussion of how to retrieve single rows using embedded SQL commands and how to use cursors to retrieve multiple rows. Chapter 8 also includes a section showing some techniques for using SQL in Visual Basic (Access). The chapter concludes with a discussion of triggers.

Appendix A: SQL Reference

Appendix A includes a command reference for all the major SQL clauses and operators that are featured in the chapters. Students can use Appendix A as a quick resource when constructing commands. Each command includes a short description, a table that shows the required and optional clauses and operators, and an example and its results. It also contains a reference to the pages in the text where the command is covered.

Appendix B: How Do I Reference

Appendix B provides students with an opportunity to ask a question, such as "How do I delete rows?", and to identify the appropriate section in Appendix A to use to find the answer. Appendix B is extremely valuable when students know what task they want to accomplish but can't remember the exact SQL command they need.

Appendix C: Answers to Odd-Numbered Review Questions

Answers to the odd-numbered Review Questions in each chapter appear in this appendix so students can make sure that they are completing the Review Questions correctly.

GENERAL NOTES TO THE STUDENT

You can download the databases used in this text from *www.cengage.com*. The Access data files for this book include three Access 2003 databases (Premiere Products.mdb, Henry Books.mdb, and Alexamara Marina Group.mdb), which you can open in Access 2000, 2002, or 2003; and three Access 2007 databases (Premiere Products.accdb, Henry Books.accdb, and Alexamara Marina Group.accdb), which you can open in Access 2007.

The data files also include script files for Oracle and SQL Server that you can use to create or drop the Premiere Products, Henry Books, and Alexamara Marina Group databases.

The script files saved in the Oracle folder have the following functions:

Oracle-Alexamara.sql: Creates all the tables in the Alexamara Marina Group database and adds all the data. Run this script file to create the Alexamara Marina Group database. (*Note:* This script file assumes you have not previously created any of the tables in the database. If you have created any of the tables, you should run the Oracle-DropAlexamara.sql script prior to running the Oracle-Alexamara.sql script.)

Oracle-Henry.sql: Creates all the tables in the Henry Books database and adds all the data. Run this script file to create the Henry Books database. (*Note:* This script file assumes you have not previously created any of the tables in the database. If you have created any of the tables, you should run the Oracle-DropHenry.sql script prior to running the Oracle-Henry.sql script.)

Oracle-Premiere.sql: Creates all the tables in the Premiere Products database and adds all the data. Run this script file to create the Premiere Products database. (*Note:* This script file assumes you have not previously created any of the tables in the database. If you have created any of the tables, you should run the Oracle-DropPremiere.sql script prior to running the Oracle-Premiere.sql script.)

Oracle-DropAlexamara.sql: Drops (deletes) all the tables and data in the Alexamara Marina Group database.

Oracle-DropHenry.sql: Drops (deletes) all the tables and data in the Henry Books database.

Oracle-DropPremiere.sql: Drops (deletes) all the tables and data in the Premiere Products database.

The script files saved in the SQL Server folder have the following functions:

SQLServer-Alexamara.sql: Creates all the tables in the Alexamara Marina Group database and adds all the data. Run this script file to create the Alexamara Marina Group database. (*Note:* This script file assumes you have not previously created any of the tables in the database. If you have created any of the tables, you should run the SQLServer-DropAlexamara.sql script prior to running the SQLServer-Alexamara.sql script.)

SQLServer-Henry.sql: Creates all the tables in the Henry Books database and adds all the data. Run this script file to create the Henry Books database. (*Note:* This script file assumes you have not previously created any of the tables in the database. If you have created any of the tables, you should run the SQLServer-DropHenry.sql script prior to running the SQLServer-Henry.sql script.)

SQLServer-Premiere.sql: Creates all the tables in the Premiere Products database and adds all the data. Run this script file to create the Premiere Products database. (*Note:* This script file assumes you have not previously created any of the tables in the database. If you have created any of the tables, you should run the SQLServer-DropPremiere.sql script prior to running the SQLServer-Premiere.sql script.)

SQLServer-DropAlexamara.sql: Drops (deletes) all the tables and data in the Alexamara Marina Group database.

SQLServer-DropHenry.sql: Drops (deletes) all the tables and data in the Henry Books database.

SQLServer-DropPremiere.sql: Drops (deletes) all the tables and data in the Premiere Products database.

For details on running script files in Oracle or SQL Server, check with your instructor. You can also refer to Chapter 3 in the text for information about creating and using scripts.

For information about downloading the Oracle Database 10g Express Edition software, please visit the Oracle Web site. For information about SQL Server 2005, please visit the Microsoft Web site. Information about Microsoft Access is also available at the Microsoft Web site.

Embedded Questions

In many places, you'll find Q&A sections to ensure that you understand some crucial material before you proceed. In some cases, the questions are designed to give you the chance to consider some special concept in advance of its actual presentation. In all cases, the answer to each question appears immediately after the question. You can simply read the question and its answer, but you will benefit from taking time to determine the answer to the question before checking your answer against the one given in the text.

End-of-Chapter Material

The end-of-chapter material consists of a Summary, a Key Terms list, Review Questions, and exercises for the Premiere Products, Henry Books, and Alexamara Marina Group databases. The Summary briefly describes the material covered in the chapter. The Review Questions require you to recall and apply the important material in the chapter. The answers to the odd-numbered Review Questions appear in Appendix C so you can check your progress. The Premiere Products, Henry Books, and Alexamara Marina Group exercises test your knowledge of the chapter material; your instructor will assign one or more of these exercises for you to complete.

ACKNOWLEDGMENTS

We would like to acknowledge several individuals for their contributions in the preparation of this text. We appreciate the efforts of the following individuals who reviewed the manuscript and made many helpful suggestions: Vickee Stedham, St. Petersburg College; Bill Kloepfer, Golden Gate University; Georgia Brown, Northern Illinois University; Gary Savard, Champlain College; Stephen Cerovski, Coleman College; Ricardo Herrera, Vanier College and Concordia University; Eugenia Fernandez, Indiana University-Purdue University Indianapolis; Danny Yakimchuk, University College of Cape Breton; Paul Leidig, Grand Valley State University; Misty Vermaat, Purdue University Calumet; Lorna Bowen St. George, Old Dominion University; and George Federman, Santa Barbara Community College.

The efforts of the following members of the staff at Course Technology have been invaluable and have made this text possible: Charles McCormick, Senior Acquisitions Editor; Kate Hennessy, Product Manager; Matt Hutchinson, Content Project Manager; Marisa Taylor, Project Manager; and GreenPen Quality Assurance testers.

We have once again had the great pleasure to work with an absolutely amazing Developmental Editor, Jessica Evans, on several books. Thanks for all your efforts, Jess. You're the best! We've said it before, but it is just as true as ever!

CHAPTER 1

INTRODUCTION TO PREMIERE PRODUCTS, HENRY BOOKS, AND ALEXAMARA MARINA GROUP

LEARNING OBJECTIVES

Objectives

- Introduce Premiere Products, a company whose database is used as the basis for many of the examples throughout the text
- Introduce Henry Books, a company whose database is used as a case that runs throughout the text
- Introduce Alexamara Marina Group, a company whose database is used as an additional case that runs throughout the text

INTRODUCTION

In this chapter, you will examine the database requirements of Premiere Products, a company that will be used in the examples throughout the text. Then you will examine the database requirements for Henry Books and Alexamara Marina Group, whose databases are featured in the exercises that appear at the end of each chapter.

WHAT IS A DATABASE?

Throughout this text, you will work with databases for three organizations: Premiere Products, Henry Books, and Alexamara Marina Group. A **database** is a structure that contains different categories of information and the relationships between these categories. The Premiere Products database, for example, contains information about categories such as sales representatives (sales reps), customers, orders, and parts. The Henry Books database contains information about categories such as books, publishers, authors, and branches. The Alexamara Marina Group database contains information about categories such as marinas, slips and the boats in them, service categories, and service requests.

Each database also contains relationships between categories. For example, the Premiere Products database contains information that relates sales reps to the customers they represent and customers to the orders they have placed. The Henry Books database contains information that relates publishers to the books they publish and authors to the books they have written. The Alexamara Marina Group database contains information that relates the boats in the slips at the marina to the owners of the boats.

As you work through the chapters in this text, you will learn more about these databases and how to view and update the information they contain. As you read each chapter, you will see examples from the Premiere Products database. At the end of each chapter, your instructor might assign the exercises for the Premiere Products, Henry Books, or Alexamara Marina Group databases.

THE PREMIERE PRODUCTS DATABASE

The management of Premiere Products, a distributor of appliances, housewares, and sporting goods, has determined that the company's recent growth no longer makes it feasible to maintain customer, order, and inventory data using its manual systems. With the data stored in a database, management will be able to ensure that the data is current and more accurate than in the present manual systems. In addition, managers will be able to obtain answers to their questions concerning the data in the database easily and quickly, with the option of producing a variety of useful reports.

Management has determined that Premiere Products must maintain the following information about its sales reps, customers, and parts inventory in the new database:

- The number, last name, first name, address, total commission, and commission rate for each sales rep
- The customer number, name, address, current balance, and credit limit for each customer, as well as the number of the sales rep who represents the customer
- The part number, description, number of units on hand, item class, number of the warehouse where the item is stored, and unit price for each part in inventory

Premiere Products also must store information about orders. Figure 1-1 shows a sample order.

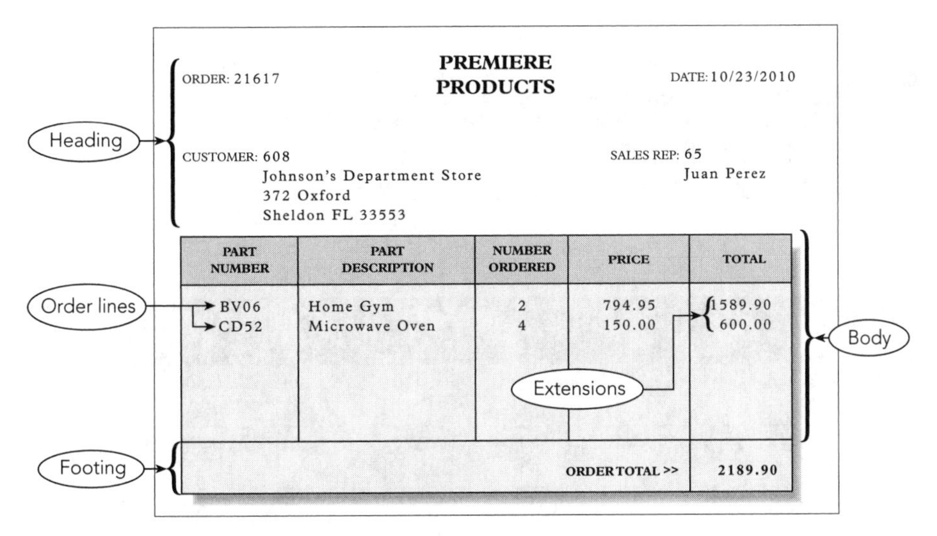

FIGURE 1-1 Sample order

The sample order shown in Figure 1-1 has three sections:

- The heading (top) of the order contains the company name; the order number and date; the customer's number, name, and address; and the sales rep's number and name.
- The body of the order contains one or more order lines, sometimes called line items. Each order line contains a part number, a part description, the number of units of the part ordered, and the quoted price for the part. Each order line also contains a total, usually called an extension, which is the result of multiplying the number ordered by the quoted price.
- Finally, the footing (bottom) of the order contains the order total.

Premiere Products also must store the following items in the database for each customer's order:

- For each order, the database must store the order number, the date the order was placed, and the number of the customer that placed the order. The customer's name and address and the number of the sales rep who represents the customer are stored with the customer information. The name of the sales rep is stored with the sales rep information.
- For each order, the database must store the order number, the part number, the number of units ordered, and the quoted price for each order line. The part description is stored with the information about parts. The result of multiplying the number of units ordered by the quoted price is not stored because the database can calculate it when needed.
- The overall order total is not stored. Instead, the database calculates the total whenever an order is printed or displayed on the screen.

Figure 1-2 shows sample data for Premiere Products.

REP

REP_ NUM	LAST_ NAME	FIRST_ NAME	STREET	CITY	STATE	ZIP	COMMISSION	RATE
20	Kaiser	Valerie	624 Randall	Grove	FL	33321	$20,542.50	0.05
35	Hull	Richard	532 Jackson	Sheldon	FL	33553	$39,216.00	0.07
65	Perez	Juan	1626 Taylor	Fillmore	FL	33336	$23,487.00	0.05

CUSTOMER

CUSTOMER_ NUM	CUSTOMER_ NAME	STREET	CITY	STATE	ZIP	BALANCE	CREDIT_ LIMIT	REP_ NUM
148	Al's Appliance and Sport	2837 Greenway	Fillmore	FL	33336	$6,550.00	$7,500.00	20
282	Brookings Direct	3827 Devon	Grove	FL	33321	$431.50	$10,000.00	35
356	Ferguson's	382 Wildwood	Northfield	FL	33146	$5,785.00	$7,500.00	65
408	The Everything Shop	1828 Raven	Crystal	FL	33503	$5,285.25	$5,000.00	35
462	Bargains Galore	3829 Central	Grove	FL	33321	$3,412.00	$10,000.00	65
524	Kline's	838 Ridgeland	Fillmore	FL	33336	$12,762.00	$15,000.00	20
608	Johnson's Department Store	372 Oxford	Sheldon	FL	33553	$2,106.00	$10,000.00	65
687	Lee's Sport and Appliance	282 Evergreen	Altonville	FL	32543	$2,851.00	$5,000.00	35
725	Deerfield's Four Seasons	282 Columbia	Sheldon	FL	33553	$248.00	$7,500.00	35
842	All Season	28 Lakeview	Grove	FL	33321	$8,221.00	$7,500.00	20

FIGURE 1-2 Sample data for Premiere Products

ORDERS

ORDER_ NUM	ORDER_ DATE	CUSTOMER_ NUM
21608	10/20/2010	148
21610	10/20/2010	356
21613	10/21/2010	408
21614	10/21/2010	282
21617	10/23/2010	608
21619	10/23/2010	148
21623	10/23/2010	608

ORDER_LINE

ORDER_NUM	PART_NUM	NUM_ORDERED	QUOTED_PRICE
21608	AT94	11	$21.95
21610	DR93	1	$495.00
21610	DW11	1	$399.99
21613	KL62	4	$329.95
21614	KT03	2	$595.00
21617	BV06	2	$794.95
21617	CD52	4	$150.00
21619	DR93	1	$495.00
21623	KV29	2	$1,290.00

PART

PART_NUM	DESCRIPTION	ON_HAND	CLASS	WAREHOUSE	PRICE
AT94	Iron	50	HW	3	$24.95
BV06	Home Gym	45	SG	2	$794.95
CD52	Microwave Oven	32	AP	1	$165.00
DL71	Cordless Drill	21	HW	3	$129.95
DR93	Gas Range	8	AP	2	$495.00
DW11	Washer	12	AP	3	$399.99
FD21	Stand Mixer	22	HW	3	$159.95
KL62	Dryer	12	AP	1	$349.95
KT03	Dishwasher	8	AP	3	$595.00
KV29	Treadmill	9	SG	2	$1,390.00

FIGURE 1-2 Sample data for Premiere Products (continued)

In the REP table, you see that there are three sales reps, whose numbers are 20, 35, and 65. The name of sales rep 20 is Valerie Kaiser. Her street address is 624 Randall. She lives in Grove, Florida, and her zip code is 33321. Her total commission is $20,542.50, and her commission rate is five percent (0.05).

In the CUSTOMER table, 10 Premiere Products customers are identified with the numbers 148, 282, 356, 408, 462, 524, 608, 687, 725, and 842. The name of customer number 148 is Al's Appliance and Sport. This customer's address is 2837 Greenway in Fillmore, Florida, with a zip code of 33336. The customer's current balance is $6,550.00, and its credit limit is $7,500.00. The number 20 in the REP_NUM column indicates that Al's Appliance and Sport is represented by sales rep 20 (Valerie Kaiser).

Skipping to the table named PART, you see that there are 10 parts, whose part numbers are AT94, BV06, CD52, DL71, DR93, DW11, FD21, KL62, KT03, and KV29. Part AT94 is an iron, and the company has 50 units of this part on hand. Irons are in item class HW (housewares) and are stored in warehouse 3. The price of an iron is $24.95. Other item classes are AP (appliances) and SG (sporting goods).

Moving back to the table named ORDERS, you see that there are seven orders, which are identified with the numbers 21608, 21610, 21613, 21614, 21617, 21619, and 21623. Order number 21608 was placed on October 20, 2010, by customer 148 (Al's Appliance and Sport).

NOTE

In some database systems, the word *order* has a special purpose. Having a table named ORDER could cause problems in such systems. For this reason, Premiere Products uses the table name ORDERS instead of ORDER.

The table named ORDER_LINE might seem strange at first glance. Why do you need a separate table for the order lines? Could they be included in the ORDERS table? The answer is technically yes. You could structure the table named ORDERS as shown in Figure 1-3. Notice that this table contains the same orders as shown in Figure 1-2, with the same dates and customer numbers. In addition, each table row in Figure 1-3 contains all the order lines for a given order. Examining the fifth row, for example, you see that order 21617 has two order lines. One of these order lines is for two BV06 parts at $794.95 each, and the other order line is for four CD52 parts at $150.00 each.

ORDERS

ORDER_NUM	ORDER_DATE	CUSTOMER_NUM	PART_NUM	NUM_ORDERED	QUOTED_PRICE
21608	10/20/2010	148	AT94	11	$21.95
21610	10/20/2010	356	DR93 DW11	1 1	$495.00 $399.99
21613	10/21/2010	408	KL62	4	$329.95
21614	10/21/2010	282	KT03	2	$595.00
21617	10/23/2010	608	BV06 CD52	2 4	$794.95 $150.00
21619	10/23/2010	148	DR93	1	$495.00
21623	10/23/2010	608	KV29	2	$1,290.00

FIGURE 1-3 Alternative ORDERS table structure

Q & A

Question: How is the information from Figure 1-2 represented in Figure 1-3?
Answer: Examine the ORDER_LINE table shown in Figure 1-2 and note the sixth and seventh rows. The sixth row indicates that there is an order line on order 21617 for two BV06 parts at $794.95 each. The seventh row indicates that there is an order line on order 21617 for four CD52 parts at $150.00 each. Thus, the information that you find in Figure 1-3 is represented in Figure 1-2 in two separate rows rather than in one row.

It might seem inefficient to use two rows to store information that could be represented in one row. There is a problem, however, with the arrangement shown in Figure

1-3—the table is more complicated. In Figure 1-2, there is a single entry at each location in the table. In Figure 1-3, some of the individual positions within the table contain multiple entries, making it difficult to track the information between columns. In the row for order number 21617, for example, it is crucial to know that the BV06 corresponds to the 2 in the NUM_ORDERED column (not the 4) and that it corresponds to the $794.95 in the QUOTED_PRICE column (not the $150.00). In addition, a more complex table raises practical issues, such as:

- How much room do you allow for these multiple entries?
- What happens when an order has more order lines than you have allowed room for?
- For a given part, how do you determine which orders contain order lines for that part?

Although none of these problems is unsolvable, they do add a level of complexity that is not present in the arrangement shown in Figure 1-2. In Figure 1-2, there are no multiple entries to worry about, it does not matter how many order lines exist for any order, and finding every order that contains an order line for a given part is easy (just look for all order lines with the given part number in the PART_NUM column). In general, this simpler structure is preferable, and that is why order lines appear in a separate table.

To test your understanding of the Premiere Products data, use Figure 1-2 to answer the following questions.

Q & A

Question: What are the numbers of the customers represented by Valerie Kaiser?
Answer: 148, 524, and 842. (Look up the REP_NUM value of Valerie Kaiser in the REP table and obtain the number 20. Then find all customers in the CUSTOMER table that have the number 20 in the REP_NUM column.)

Q & A

Question: What is the name of the customer that placed order 21610, and what is the name of the rep who represents this customer?
Answer: Ferguson's is the customer; Juan Perez is the sales rep. (Look up the CUSTOMER_NUM value in the ORDERS table for order number 21610 and obtain the number 356. Then find the customer in the CUSTOMER table with the CUSTOMER_NUM value of 356. Using the REP_NUM value, which is 65, find the name of the rep in the REP table.)

Q & A

Question: List all parts that appear in order 21610. For each part, give the description, number ordered, and quoted price.
Answer: Part number: DR93; part description: Gas Range; number ordered: 1; and quoted price: $495.00. Also, part number: DW11; part description: Washer; number ordered: 1; and quoted price: $399.99. (Look up each ORDER_LINE table row in which the order number is 21610. Each of these rows contains a part number, the number ordered, and the quoted price. Use the part number to look up the corresponding part description in the PART table.)

Q & A

Question: Why is the QUOTED_PRICE column part of the ORDER_LINE table? Can't you just use the part number and look up the price in the PART table?
Answer: If the QUOTED_PRICE column did not appear in the ORDER_LINE table, you would need to obtain the price for a part on an order line by looking up the price in the PART table. Although this approach is reasonable, it prevents Premiere Products from charging different prices to different customers for the same part. Because Premiere Products wants the flexibility to quote and charge different prices to different customers, the QUOTED_PRICE column is included in the ORDER_LINE table. If you examine the ORDER_LINE table, you will see cases in which the quoted price matches the actual price in the PART table and cases in which it differs. For example, in order number 21608, Al's Appliance and Sport bought 11 irons, and Premiere Products charged only $21.95 per iron, rather than the regular price of $24.95.

THE HENRY BOOKS DATABASE

Ray Henry is the owner of a bookstore chain named Henry Books. Like the management of Premiere Products, Ray has decided to store his data in a database. He wants to achieve the same benefits; that is, he wants to ensure that his data is current and accurate. In addition, he wants to be able to ask questions concerning the data and to obtain answers to these questions easily and quickly.

In running his chain of bookstores, Ray gathers and organizes information about branches, publishers, authors, and books. Figure 1-4 shows sample branch and publisher data for Henry Books. Each branch has a number that uniquely identifies the branch. In addition, Ray tracks the branch's name, location, and number of employees. Each publisher has a code that uniquely identifies the publisher. In addition, Ray tracks the publisher's name and city.

BRANCH

BRANCH_NUM	BRANCH_NAME	BRANCH_LOCATION	NUM_EMPLOYEES
1	Henry Downtown	16 Riverview	10
2	Henry On The Hill	1289 Bedford	6
3	Henry Brentwood	Brentwood Mall	15
4	Henry Eastshore	Eastshore Mall	9

PUBLISHER

PUBLISHER_CODE	PUBLISHER_NAME	CITY
AH	Arkham House	Sauk City WI
AP	Arcade Publishing	New York
BA	Basic Books	Boulder CO
BP	Berkley Publishing	Boston
BY	Back Bay Books	New York
CT	Course Technology	Boston
FA	Fawcett Books	New York
FS	Farrar Straus and Giroux	New York
HC	HarperCollins Publishers	New York
JP	Jove Publications	New York
JT	Jeremy P. Tarcher	Los Angeles
LB	Lb Books	New York
MP	McPherson and Co.	Kingston
PE	Penguin USA	New York
PL	Plume	New York
PU	Putnam Publishing Group	New York
RH	Random House	New York
SB	Schoken Books	New York
SC	Scribner	New York
SS	Simon and Schuster	New York
ST	Scholastic Trade	New York
TA	Taunton Press	Newtown CT
TB	Tor Books	New York
TH	Thames and Hudson	New York
TO	Touchstone Books	Westport CT
VB	Vintage Books	New York
WN	W.W. Norton	New York
WP	Westview Press	Boulder CO

FIGURE 1-4 Sample branch and publisher data for Henry Books

Figure 1-5 shows sample author data for Henry Books. Each author has a number that uniquely identifies the author. In addition, Ray records each author's last and first names.

AUTHOR

AUTHOR_NUM	AUTHOR_LAST	AUTHOR_FIRST
1	Morrison	Toni
2	Solotaroff	Paul
3	Vintage	Vernor
4	Francis	Dick
5	Straub	Peter
6	King	Stephen
7	Pratt	Philip
8	Chase	Truddi
9	Collins	Bradley
10	Heller	Joseph
11	Wills	Gary
12	Hofstadter	Douglas R.
13	Lee	Harper
14	Ambrose	Stephen E.
15	Rowling	J.K.
16	Salinger	J.D.
17	Heaney	Seamus
18	Camus	Albert
19	Collins, Jr.	Bradley
20	Steinbeck	John
21	Castelman	Riva
22	Owen	Barbara
23	O'Rourke	Randy
24	Kidder	Tracy
25	Schleining	Lon

FIGURE 1-5 Sample author data for Henry Books

Figure 1-6 shows sample book data for Henry Books. Each book has a code that uniquely identifies the book. For each book, Ray also tracks the title, publisher, book type, price, and whether the book is a paperback.

BOOK

BOOK_CODE	TITLE	PUBLISHER_ CODE	TYPE	PRICE	PAPERBACK
0180	A Deepness in the Sky	TB	SFI	$7.19	Y
0189	Magic Terror	FA	HOR	$7.99	Y
0200	The Stranger	VB	FIC	$8.00	Y
0378	Venice	SS	ART	$24.50	N
079X	Second Wind	PU	MYS	$24.95	N
0808	The Edge	JP	MYS	$6.99	Y
1351	Dreamcatcher: A Nvel	SC	HOR	$19.60	N
1382	Treasure Chests	TA	ART	$24.46	N
138X	Beloved	PL	FIC	$12.95	Y
2226	Harry Potter and the Prisoner of Azkaban	ST	SFI	$13.96	N
2281	Van Gogh and Gauguin	WP	ART	$21.00	N
2766	Of Mice and Men	PE	FIC	$6.95	Y
2908	Electric Light	FS	POE	$14.00	N
3350	Group: Six People in Search of a Life	BP	PSY	$10.40	Y
3743	Nine Stories	LB	FIC	$5.99	Y
3906	The Soul of a New Machine	BY	SCI	$11.16	Y
5163	Travels with Charley	PE	TRA	$7.95	Y
5790	Catch-22	SC	FIC	$12.00	Y
6128	Jazz	PL	FIC	$12.95	Y
6328	Band of Brothers	TO	HIS	$9.60	Y
669X	A Guide to SQL	CT	CMP	$37.95	Y
6908	Franny and Zooey	LB	FIC	$5.99	Y
7405	East of Eden	PE	FIC	$12.95	Y
7443	Harry Potter and the Goblet of Fire	ST	SFI	$18.16	N
7559	The Fall	VB	FIC	$8.00	Y
8092	Godel, Escher, Bach	BA	PHI	$14.00	Y
8720	When Rabbit Howls	JP	PSY	$6.29	Y
9611	Black House	RH	HOR	$18.81	N
9627	Song of Solomon	PL	FIC	$14.00	Y
9701	The Grapes of Wrath	PE	FIC	$13.00	Y
9882	Slay Ride	JP	MYS	$6.99	Y
9883	The Catcher in the Rye	LB	FIC	$5.99	Y
9931	To Kill a Mockingbird	HC	FIC	$18.00	N

FIGURE 1-6 Sample book data for Henry Books

To check your understanding of the relationship between publishers and books, answer the following questions.

Q & A

Question: Who published *Jazz*? Which books did Jove Publications publish?
Answer: Plume published *Jazz*. In the row in the BOOK table for *Jazz* (see Figure 1-6), find the publisher code PL. Examining the PUBLISHER table (see Figure 1-4), you see that PL is the code assigned to Plume. Jove Publications published *The Edge*, *When Rabbit Howls*, and *Slay Ride*. To find the books published by Jove Publications, find its code (JP) in the PUBLISHER table. Next, find all records in the BOOK table for which the publisher code is JP.

The table named WROTE, as shown in Figure 1-7, relates books to the authors who wrote them. The SEQUENCE column indicates the order in which the authors of a particular book are listed on the cover. The table named INVENTORY in the same figure is used to indicate the number of copies of a particular book that are currently on hand at a particular branch of Henry Books. The first row, for example, indicates that there are two copies of the book with the code 0180 at branch 1.

WROTE

BOOK_CODE	AUTHOR_NUM	SEQUENCE
0180	3	1
0189	5	1
0200	18	1
0378	11	1
079X	4	1
0808	4	1
1351	6	1
1382	23	2
1382	25	1
138X	1	1
2226	15	1
2281	9	2
2281	19	1
2766	20	1
2908	17	1
3350	2	1
3743	16	1

INVENTORY

BOOK_CODE	BRANCH_NUM	ON_HAND
0180	1	2
0189	2	2
0200	1	1
0200	2	3
0378	3	2
079X	2	1
079X	3	2
079X	4	3
0808	2	1
1351	2	4
1351	3	2
1382	2	1
138X	2	3
2226	1	3
2226	3	2
2226	4	1
2281	4	3

FIGURE 1-7 Sample data that relates books to authors and books to branches for Henry Books

WROTE

BOOK_CODE	AUTHOR_NUM	SEQUENCE
3906	24	1
5163	20	1
5790	10	1
6128	1	1
6328	14	1
669X	7	1
6908	16	1
7405	20	1
7443	15	1
7559	18	1
8092	12	1
8720	8	1
9611	5	2
9611	6	1
9627	1	1
9701	20	1
9882	4	1
9883	16	1
9931	13	1

INVENTORY

BOOK_CODE	BRANCH_NUM	ON_HAND
2766	3	2
2908	1	3
2908	4	1
3350	1	2
3743	2	1
3906	2	1
3906	3	2
5163	1	1
5790	4	2
6128	2	4
6128	3	3
6328	2	2
669X	1	1
6908	2	2
7405	3	2
7443	4	1
7559	2	2
8092	3	1
8720	1	3
9611	1	2
9627	3	5
9627	4	2
9701	1	2
9701	2	1
9701	3	3
9701	4	2
9882	3	3
9883	2	3
9883	4	2
9931	1	2

FIGURE 1-7 Sample data that relates books to authors and books to branches for Henry Books (continued)

To check your understanding of the relationship between authors and books, answer the following questions.

Q & A

Question: Who wrote *Black House*? (Make sure to list the authors in the correct order.) Which books did Toni Morrison write?
Answer: Stephen King and Peter Straub wrote *Black House*. First examine the BOOK table (see Figure 1-6) to find the book code for *Black House* (9611). Next, look for all rows in the WROTE table in which the book code is 9611. There are two such rows. In one row, the author number is 5, and in the other, it is 6. Then, look in the AUTHOR table to find the authors who have been assigned the numbers 5 and 6. The answers are Peter Straub (5) and Stephen King (6). The sequence number for author number 5 is 2, and the sequence number for author number 6 is 1. Thus, listing the authors in the proper order results in Stephen King and Peter Straub.

Toni Morrison wrote *Beloved*, *Jazz*, and *Song of Solomon*. To find the books written by Toni Morrison, look up her author number (1) in the AUTHOR table. Then look for all rows in the WROTE table for which the author number is 1. There are three such rows. The corresponding book codes are 138X, 6128, and 9627. Looking up these codes in the BOOK table, you find that Toni Morrison wrote *Beloved*, *Jazz*, and *Song of Solomon*.

Q & A

Question: A customer in branch 1 wants to purchase *The Soul of a New Machine*. Is this book currently in stock at branch 1?
Answer: No. Looking up the code for *The Soul of a New Machine* in the BOOK table, you find it is 3906. To find out how many copies are in stock at branch 1, look for a row in the INVENTORY table with 3906 in the BOOK_CODE column and 1 in the BRANCH_NUM column. Because there is no such row, branch 1 doesn't have any copies of *The Soul of a New Machine*.

Q & A

Question: You would like to obtain a copy of *The Soul of a New Machine* for this customer. Which other branches currently have this book in stock, and how many copies does each branch have?
Answer: Branch 2 has one copy, and branch 3 has two copies. You already know that the code for *The Soul of a New Machine* is 3906. (If you did not know the book code, you would look it up in the BOOK table.) To find out which branches currently have copies, look for rows in the INVENTORY table with 3906 in the BOOK_CODE column. There are two such rows. The first row indicates that branch 2 currently has one copy. The second row indicates that branch 3 currently has two copies.

THE ALEXAMARA MARINA GROUP DATABASE

Alexamara Marina Group offers in-water boat storage to owners by providing boat slips that owners can rent on an annual basis. Alexamara owns two marinas: Alexamara East and Alexamara Central. Each marina has several boat slips available. Alexamara also provides a variety of boat repair and maintenance services to the boat owners who rent the slips. Alexamara stores the data it needs to manage its operations in a relational database containing the tables described in the following section.

Alexamara stores information about its two marinas in the MARINA table shown in Figure 1-8. A marina number uniquely identifies each marina. The table also includes the marina name, street address, city, state, and zip code.

MARINA

MARINA_NUM	NAME	ADDRESS	CITY	STATE	ZIP
1	Alexamara East	108 2nd Ave.	Brinman	FL	32273
2	Alexamara Central	283 Branston	W. Brinman	FL	32274

FIGURE 1-8 Sample marina data for Alexamara Marina Group

Alexamara stores information about the boat owners to whom it rents slips in the OWNER table shown in Figure 1-9. An owner number that consists of two uppercase letters followed by a two-digit number uniquely identifies each owner. For each owner, the table also includes the last name, first name, address, city, state, and zip code.

OWNER

OWNER_ NUM	LAST_ NAME	FIRST_NAME	ADDRESS	CITY	STATE	ZIP
AD57	Adney	Bruce and Jean	208 Citrus	Bowton	FL	31313
AN75	Anderson	Bill	18 Wilcox	Glander Bay	FL	31044
BL72	Blake	Mary	2672 Commodore	Bowton	FL	31313
EL25	Elend	Sandy and Bill	462 Riverside	Rivard	FL	31062
FE82	Feenstra	Daniel	7822 Coventry	Kaleva	FL	32521
JU92	Juarez	Maria	8922 Oak	Rivard	FL	31062
KE22	Kelly	Alyssa	5271 Waters	Bowton	FL	31313
NO27	Norton	Peter	2811 Lakewood	Lewiston	FL	32765
SM72	Smeltz	Becky and Dave	922 Garland	Glander Bay	FL	31044
TR72	Trent	Ashton	922 Crest	Bay Shores	FL	30992

FIGURE 1-9 Sample owner data for Alexamara Marina Group

Each marina contains slips that are identified by slip numbers. Marina 1 (Alexamara East) has two sections (A and B) and slips are numbered within each section. Thus, slip numbers at marina 1 consist of the letter A or B followed by a number (for example, A3 or B2). At marina 2 (Alexamara Central), a number (1, 2, 3) identifies each slip.

Information about the slips in the marinas is contained in the MARINA_SLIP table shown in Figure 1-10. Each row in the table contains a slip ID that identifies the particular slip. The table also contains the marina number and slip number, the length of the slip (in feet), the annual rental fee, the name of the boat currently occupying the slip, the type of boat, and the boat owner's number.

MARINA_SLIP

SLIP_ID	MARINA_NUM	SLIP_NUM	LENGTH	RENTAL_FEE	BOAT_NAME	BOAT_TYPE	OWNER_NUM
1	1	A1	40	$3,800.00	Anderson II	Sprite 4000	AN75
2	1	A2	40	$3,800.00	Our Toy	Ray 4025	EL25
3	1	A3	40	$3,600.00	Escape	Sprite 4000	KE22
4	1	B1	30	$2,400.00	Gypsy	Dolphin 28	JU92
5	1	B2	30	$2,600.00	Anderson III	Sprite 3000	AN75
6	2	1	25	$1,800.00	Bravo	Dolphin 25	AD57
7	2	2	25	$1,800.00	Chinook	Dolphin 22	FE82
8	2	3	25	$2,000.00	Listy	Dolphin 25	SM72
9	2	4	30	$2,500.00	Mermaid	Dolphin 28	BL72
10	2	5	40	$4,200.00	Axxon II	Dolphin 40	NO27
11	2	6	40	$4,200.00	Karvel	Ray 4025	TR72

FIGURE 1-10 Sample data about slips at Alexamara Marina Group

Alexamara provides boat maintenance service for owners at its two marinas. The types of service provided are stored in the SERVICE_CATEGORY table shown in Figure 1-11. A category number uniquely identifies each service that Alexamara performs. The table also contains a description of the category.

SERVICE_CATEGORY

CATEGORY_NUM	CATEGORY_DESCRIPTION
1	Routine engine maintenance
2	Engine repair
3	Air conditioning
4	Electrical systems
5	Fiberglass repair
6	Canvas installation
7	Canvas repair
8	Electronic systems (radar, GPS, autopilots, etc.)

FIGURE 1-11 Sample data about service categories at Alexamara Marina Group

Information about the services requested by owners is stored in the SERVICE_REQUEST table shown in Figure 1-12. Each row in the table contains a service ID that identifies each service request. The slip ID identifies the location (marina number and slip number) of the boat to be serviced. For example, the slip ID on the second row is 5. As indicated in the MARINA_SLIP table in Figure 1-10, the slip ID 5 identifies the boat in marina 1 and slip number B2.

The SERVICE_REQUEST table also contains the category number of the service to be performed, plus a description of the specific service to be performed, and a description of the current status of the service. It also contains the estimated number of hours required to complete the service. For completed jobs, the table contains the actual number of hours it took to complete the service. If another appointment is required to complete additional service, the appointment date appears in the NEXT_SERVICE_DATE column.

SERVICE_REQUEST

SERVICE_ ID	SLIP_ ID	CATEGORY_ NUM	DESCRIPTION	STATUS	EST_ HOURS	SPENT_ HOURS	NEXT_ SERVICE_ DATE
1	1	3	Air conditioner periodically stops with code indicating low coolant level. Diagnose and repair.	Technician has verified the problem. Air conditioning specialist has been called.	4	2	7/12/2010
2	5	4	Fuse on port motor blown on two occasions. Diagnose and repair.	Open	2	0	7/12/2010
3	4	1	Oil change and general routine maintenance (check fliud levels, clean sea strainers, etc.).	Service call has been scheduled.	1	0	7/16/2010
4	1	2	Engine oil level has been dropping drastically. Diagnose and repair.	Open	2	0	7/13/2010
5	3	5	Open pockets at base of two stantions.	Technician has completed the initial filling of the open pockets. Will complete the job after the initial fill has had sufficient time to dry.	4	2	7/13/2010
6	11	4	Electric-flush system periodically stops functioning. Diagnose and repair.	Open	3	0	
7	6	2	Engine overheating. Loss of coolant. Diagnose and repair.	Open	2	0	7/13/2010
8	6	2	Heat exchanger not operating correctly.	Technician has determined that the exchanger is faulty. New exchanger has been ordered.	4	1	7/17/2010
9	7	6	Canvas severely damaged in windstorm. Order and install new canvas.	Open	8	0	7/16/2010
10	2	8	Install new GPS and chart plotter.	Scheduled	7	0	7/17/2010

FIGURE 1-12 Sample data about service requests at Alexamara Marina Group

SERVICE_REQUEST

SERVICE_ ID	SLIP_ ID	CATEGORY_ NUM	DESCRIPTION	STATUS	EST_ HOURS	SPENT_ HOURS	NEXT_ SERVICE_ DATE
11	2	3	Air conditioning unit shuts down with HHH showing on the control panel.	Technician not able to replicate the problem. Air conditioning unit ran fine through multiple tests. Owner to notify technician if the problem recurs.	1	1	
12	4	8	Both speed and depth readings on data unit are significantly less than the owner thinks they should be.	Technician has scheduled appointment with owner to attempt to verify the problem.	2	0	7/16/2010
13	8	2	Customer describes engine as making a clattering sound.	Technician suspects problem with either propeller or shaft and has scheduled the boat to be pulled from the water for further investigation.	5	2	7/12/2010
14	7	5	Owner accident caused damage to forward portion of port side.	Technician has scheduled repair.	6	0	7/13/2010
15	11	7	Canvas leaks around zippers in heavy rain. Install overlap around zippers to prevent leaks.	Overlap has been created. Installation has been scheduled.	8	3	7/17/2010

FIGURE 1-12 Sample data about service requests at Alexamara Marina Group (continued)

The Alexamara Marina Group exercises at the end of this chapter will give you a chance to check your understanding of the data in this database.

Chapter Summary

- Premiere Products is an organization whose information requirements include sales reps, customers, parts, orders, and order lines.
- Henry Books is an organization whose information requirements include branches, publishers, authors, books, inventory, and author sequences.
- Alexamara Marina Group is an organization whose information requirements include marinas, owners, slips, service categories, and service requests.

Key Terms

database

Exercises

Premiere Products

Answer each of the following questions using the Premiere Products data shown in Figure 1-2. No computer work is required.

1. List the names of all customers that have a credit limit of $7,500 or less.
2. List the order numbers for orders placed by customer number 608 on 10/23/2010.
3. List the part number, part description, and on-hand value for each part in item class SG. (*Hint:* On-hand value is the result of multiplying the number of units on hand by the price.)
4. List the part number and part description of all parts that are in item class HW.
5. How many customers have a balance that exceeds their credit limit?
6. What is the part number, description, and price of the least expensive part in the database?
7. For each order, list the order number, order date, customer number, and customer name.
8. For each order placed on October 21, 2010, list the order number, customer number, and customer name.
9. List the sales rep number and name for every sales rep who represents at least one customer with a credit limit of $10,000.
10. For each order placed on October 21, 2010, list the order number, part number, part description, and item class for each part ordered.

Henry Books

Answer each of the following questions using the Henry Books data shown in Figures 1-4 through 1-7. No computer work is required.

1. List the name of each publisher that is located in New York.
2. List the name of each branch that has at least nine employees.
3. List the book code and title of each book that has the type FIC.
4. List the book code and title of each book that has the type FIC and that is in paperback.
5. List the book code and title of each book that has the type FIC or whose publisher code is SC.
6. List the book code and title of each book that has the type MYS and a price of less than $20.

7. Customers who are part of a special program get a 10 percent discount off regular book prices. For the first five books in the BOOK table, list the book code, title, and discounted price. (Use the PRICE column to calculate the discounted price.)
8. Find the name of each publisher containing the word *and.*
9. List the book code and title of each book that has the type FIC, MYS, or ART.
10. How many books have the type SFI?
11. Calculate the average price for books that have the type ART.
12. For each book published by Penguin USA, list the book code and title.
13. List the book code, book title, and units on hand for each book in branch number 3.

Alexamara Marina Group

Answer each of the following questions using the Alexamara Marina Group data shown in Figures 1-8 through 1-12. No computer work is required.

1. List the owner number, last name, and first name of every boat owner.
2. List the last name and first name of every owner located in Bowton.
3. List the marina number and slip number for every slip whose length is equal to or less than 30 feet.
4. List the marina number and slip number for every boat with the type Dolphin 28.
5. List the slip number for every boat with the type Dolphin 28 that is located in marina 1.
6. List the boat name for each boat located in a slip whose length is between 25 and 30 feet.
7. List the slip number for every slip in marina 1 whose annual rental fee is less than $3,000.
8. Labor is billed at the rate of $60 per hour. List the slip ID, category number, estimated hours, and estimated labor cost for every service request. To obtain the estimated labor cost, multiply the estimated hours by 60. Use the column name ESTIMATED_COST for the estimated labor cost.
9. List the marina number and slip number for all slips containing a boat with the type Sprite 4000, Sprite 3000, or Ray 4025.
10. How many Dolphin 25 boats are stored at both marinas?
11. For every boat, list the marina number, slip number, boat name, owner number, owner's first name, and owner's last name.
12. For every service request for routine engine maintenance, list the slip ID, the description, and the status.
13. For every service request for routine engine maintenance, list the slip ID, marina number, slip number, estimated hours, spent hours, owner number, and owner's last name.

CHAPTER 2

DATABASE DESIGN FUNDAMENTALS

LEARNING OBJECTIVES

Objectives

- Understand the terms *entity*, *attribute*, and *relationship*
- Understand the terms *relation* and *relational database*
- Understand functional dependence and identify when one column is functionally dependent on another
- Understand the term *primary key* and identify primary keys in tables
- Design a database to satisfy a set of requirements
- Convert an unnormalized relation to first normal form
- Convert tables from first normal form to second normal form
- Convert tables from second normal form to third normal form
- Create an entity-relationship diagram to represent the design of a database

INTRODUCTION

In Chapter 1, you reviewed the tables and columns in the Premiere Products, Henry Books, and Alexamara Marina Group databases that you will use to complete the rest of this text. The process of determining the particular tables and columns that will comprise a database is known as **database design**. In this chapter, you will learn a method for designing a database to satisfy a set of requirements. In the process, you will learn how to identify the tables and columns in the database. You also will learn how to identify the relationships between the tables.

This chapter begins by examining some important concepts related to databases. It also presents the design method using the set of requirements that Premiere Products identified to produce the appropriate database design. The chapter then examines the process of normalization, in which you identify and fix potential problems in database designs. Finally, you will learn a way of visually representing the design of a database.

DATABASE CONCEPTS

Before learning how to design a database, you need to be familiar with some important database concepts related to relational databases, which are the types of databases you examined in Chapter 1 and that you will use throughout the rest of this text. The terms entity, attribute, and relationship are important to understand when designing a database; the concepts of functional dependence and primary keys are critical when learning about the database design process.

Relational Databases

A **relational database** is a collection of tables like the ones you examined for Premiere Products in Chapter 1 and that also appear in Figure 2-1. Formally, these tables are called relations, and this is how this type of database gets its name.

REP

REP_NUM	LAST_NAME	FIRST_NAME	STREET	CITY	STATE	ZIP	COMMISSION	RATE
20	Kaiser	Valerie	624 Randall	Grove	FL	33321	$20,542.50	0.05
35	Hull	Richard	532 Jackson	Sheldon	FL	33553	$39,216.00	0.07
65	Perez	Juan	1626 Taylor	Fillmore	FL	33336	$23,487.00	0.05

CUSTOMER

CUSTOMER_NUM	CUSTOMER_NAME	STREET	CITY	STATE	ZIP	BALANCE	CREDIT_LIMIT	REP_NUM
148	Al's Appliance and Sport	2837 Greenway	Fillmore	FL	33336	$6,550.00	$7,500.00	20
282	Brookings Direct	3827 Devon	Grove	FL	33321	$431.50	$10,000.00	35
356	Ferguson's	382 Wildwood	Northfield	FL	33146	$5,785.00	$7,500.00	65
408	The Everything Shop	1828 Raven	Crystal	FL	33503	$5,285.25	$5,000.00	35
462	Bargains Galore	3829 Central	Grove	FL	33321	$3,412.00	$10,000.00	65
524	Kline's	838 Ridgeland	Fillmore	FL	33336	$12,762.00	$15,000.00	20
608	Johnson's Department Store	372 Oxford	Sheldon	FL	33553	$2,106.00	$10,000.00	65
687	Lee's Sport and Appliance	282 Evergreen	Altonville	FL	32543	$2,851.00	$5,000.00	35
725	Deerfield's Four Seasons	282 Columbia	Sheldon	FL	33553	$248.00	$7,500.00	35
842	All Season	28 Lakeview	Grove	FL	33321	$8,221.00	$7,500.00	20

ORDERS

ORDER_NUM	ORDER_DATE	CUSTOMER_NUM
21608	10/20/2010	148
21610	10/20/2010	356
21613	10/21/2010	408
21614	10/21/2010	282
21617	10/23/2010	608
21619	10/23/2010	148
21623	10/23/2010	608

ORDER_LINE

ORDER_NUM	PART_NUM	NUM_ORDERED	QUOTED_PRICE
21608	AT94	11	$21.95
21610	DR93	1	$495.00
21610	DW11	1	$399.99
21613	KL62	4	$329.95
21614	KT03	2	$595.00
21617	BV06	2	$794.95
21617	CD52	4	$150.00
21619	DR93	1	$495.00
21623	KV29	2	$1,290.00

PART

PART_NUM	DESCRIPTION	ON_HAND	CLASS	WAREHOUSE	PRICE
AT94	Iron	50	HW	3	$24.95
BV06	Home Gym	45	SG	2	$794.95
CD52	Microwave Oven	32	AP	1	$165.00
DL71	Cordless Drill	21	HW	3	$129.95
DR93	Gas Range	8	AP	2	$495.00
DW11	Washer	12	AP	3	$399.99
FD21	Stand Mixer	22	HW	3	$159.95
KL62	Dryer	12	AP	1	$349.95
KT03	Dishwasher	8	AP	3	$595.00
KV29	Treadmill	9	SG	2	$1,390.00

FIGURE 2-1 Sample data for Premiere Products

NOTE

The names of columns and tables in this text follow a common naming convention in which column names use uppercase letters and replace spaces between words with underscores (_). For example, Premiere Products uses the column named LAST_NAME to store last names and the column named CREDIT_LIMIT to store credit limits.

Entities, Attributes, and Relationships

There are some terms and concepts that are very important for you to know when working in the database environment. The terms entity, attribute, and relationship are fundamental when discussing databases. An **entity** is like a noun; it is a person, place, thing, or event. The entities of interest to Premiere Products, for example, are such things as customers, orders, and sales reps. The entities that are of interest to a school include students, faculty, and classes; a real estate agency is interested in clients, houses, and agents; and a used car dealer is interested in vehicles, customers, and manufacturers.

An **attribute** is a property of an entity. The term is used here exactly as it is used in everyday English. For the entity *person*, for example, the list of attributes might include such things as eye color and height. For Premiere Products, the attributes of interest for the entity *customer* are such things as name, address, city, and so on. For the entity *faculty* at a school, the attributes would be such things as faculty number, name, office number, phone, and so on. For the entity *vehicle* at a car dealership, the attributes are such things as the vehicle identification number, model, color, year, and so on.

A **relationship** is the association between entities. There is an association between customers and sales reps, for example, at Premiere Products. A sales rep is associated with all of his or her customers, and a customer is associated with his or her sales rep. Technically, you say that a sales rep is *related* to all of his or her customers, and a customer is *related* to his or her sales rep.

The relationship between sales reps and customers is an example of a **one-to-many relationship** because one sales rep is associated with many customers, but each customer is associated with only one sales rep. (In this type of relationship, the word *many* is used in a way that is different from everyday English; it might not always mean a large number. In this context, for example, the term *many* means that a sales rep might be associated with *any* number of customers. That is, one sales rep can be associated with zero, one, or more customers.)

How does a relational database handle entities, attributes of entities, and relationships between entities? Entities and attributes are fairly simple. Each entity has its own table. In the Premiere Products database, there is one table for sales reps, one table for customers, and so on. The attributes of an entity become the columns in the table. In the table for sales reps, for example, there is a column for the sales rep number, a column for the sales rep's first name, and so on.

What about relationships? At Premiere Products, there is a one-to-many relationship between sales reps and customers (each sales rep is related to the *many* customers that he or she represents, and each customer is related to the *one* sales rep who represents the customer). How is this relationship implemented in a relational database?

Consider Figure 2-1 again. If you want to determine the name of the sales rep who represents Brookings Direct (customer number 282), you would locate the row for Brookings Direct in the CUSTOMER table and determine that the value for REP_NUM is 35. Then you would look for the row in the REP table on which the REP_NUM is 35. The *one* rep with REP_NUM 35 is Richard Hull, who represents Brookings Direct.

On the other hand, if you want to determine the names of all the customers of the rep named Valerie Kaiser, you would locate the row for Valerie Kaiser in the REP table and determine that the value in the REP_NUM column is 20. Then you would look for all the rows in the CUSTOMER table on which the REP_NUM is 20. After identifying Valerie Kaiser's rep number, you find that the *many* customers she represents are numbered 148 (Al's Appliance and Sport), 524 (Kline's), and 842 (All Season).

You implement these relationships by having common columns in two or more tables. The REP_NUM column in the REP table and the REP_NUM column in the CUSTOMER table are used to implement the relationship between sales reps and customers. Given a sales rep, you can use these columns to determine all the customers that he or she represents; given a customer, you can use these columns to find the sales rep who represents the customer.

In this context, a relation is essentially a two-dimensional table. If you consider the tables shown in Figure 2-1, however, you can see that certain restrictions are placed on relations. Each column has a unique name, and entries within each column should "match" this column name. For example, if the column name is CREDIT_LIMIT, all entries in that column must be credit limits. Also, each row should be unique—when two rows are identical, the second row does not provide any new information. For maximum flexibility, the order of the columns and rows should be immaterial. Finally, the table's design should be as simple as possible by restricting each position to a single entry and by preventing multiple entries (also called **repeating groups**) in an individual location in the table. Figure 2-2 shows a table design that includes repeating groups.

ORDERS

ORDER_ NUM	ORDER_ DATE	CUSTOMER_ NUM	PART_ NUM	NUM_ ORDERED	QUOTED_ PRICE
21608	10/20/2010	148	AT94	11	$21.95
21610	10/20/2010	356	DR93 DW11	1 1	$495.00 $399.99
21613	10/21/2010	408	KL62	4	$329.95
21614	10/21/2010	282	KT03	2	$595.00
21617	10/23/2010	608	BV06 CD52	2 4	$12.95 $150.00
21619	10/23/2010	148	DR93	1	$495.00
21623	10/23/2010	608	KV29	2	$325.99

FIGURE 2-2 Table with repeating groups

Figure 2-3 shows a better way to represent the same information shown in Figure 2-2. In Figure 2-3, every position in the table contains a single value.

ORDERS

ORDER_ NUM	ORDER_ DATE	CUSTOMER_ NUM	PART_ NUM	NUM_ ORDERED	QUOTED_ PRICE
21608	10/20/2010	148	AT94	11	$21.95
21610	10/20/2010	356	DR93	1	$495.00
21610	10/20/2010	356	DW11	1	$399.99
21613	10/21/2010	408	KL62	4	$329.95
21614	10/21/2010	282	KT03	2	$595.00
21617	10/23/2010	608	BV06	2	$12.95
21617	10/23/2010	608	CD52	4	$150.00
21619	10/23/2010	148	DR93	1	$495.00
21623	10/23/2010	608	KV29	2	$325.99

FIGURE 2-3 ORDERS data without repeating groups

When you remove the repeating groups from Figure 2-2, all of the rows in Figure 2-3 are single-valued. This structure is formally called a relation. A **relation** is a two-dimensional table in which the entries in the table are single-valued (each location in the table contains a single entry), each column has a distinct name, all values in the column match this name, the order of the rows and columns is immaterial, and each row contains unique values. A relational database is a collection of relations.

NOTE

Rows in a table (relation) are also called **records** or **tuples**. Columns in a table (relation) are also called **fields** or attributes. This text uses the terms tables, columns, and rows unless the more formal terms of relation, attributes, and tuples are necessary for clarity.

There is a commonly accepted shorthand representation to show the tables and columns in a relational database: for each table, you write the name of the table and then within parentheses list all of the columns in the table. In this representation, each table appears on its own line. Using this method, you represent the Premiere Products database as follows:

```
REP (REP_NUM, LAST_NAME, FIRST_NAME, STREET,
     CITY, STATE, ZIP, COMMISSION, RATE)
CUSTOMER (CUSTOMER_NUM, CUSTOMER_NAME, STREET,
     CITY, STATE, ZIP, BALANCE, CREDIT_LIMIT,
     REP_NUM)
ORDERS (ORDER_NUM, ORDER_DATE, CUSTOMER_NUM)
ORDER_LINE (ORDER_NUM, PART_NUM, NUM_ORDERED,
     QUOTED_PRICE)
PART (PART_NUM, DESCRIPTION, ON_HAND, CLASS,
     WAREHOUSE, PRICE)
```

Notice that some tables contain columns with duplicate names. For example, the REP_NUM column appears in both the REP table *and* the CUSTOMER table. Suppose a situation existed wherein someone (or the DBMS) might confuse the two columns. For example, if you write REP_NUM, it is not clear which REP_NUM column you want to use. You need a mechanism for indicating the REP_NUM column to which you are referring. One common approach to solving this problem is to write both the table name and the column name, separated by a period. Thus, you would reference the REP_NUM column in the CUSTOMER table as CUSTOMER.REP_NUM, and the REP_NUM column in the REP table as REP.REP_NUM. Technically, when you reference columns in this format, you say that you **qualify** the names. It is *always* acceptable to qualify column names, even when there is no potential for confusion. If confusion might arise, however, it is *essential* to qualify column names.

FUNCTIONAL DEPENDENCE

The concept of functional dependence is crucial to understanding the rest of the material in this chapter. Functional dependence is a formal name for what is basically a simple idea. To illustrate functional dependence, suppose the REP table for Premiere Products is structured as shown in Figure 2-4. The only difference between the REP table shown in Figure 2-4 and the one shown in Figure 2-1 is the addition of an extra column named PAY_CLASS.

REP

REP_ NUM	LAST_ NAME	FIRST_ NAME	STREET	CITY	STATE	ZIP	COMMISSION	PAY_ CLASS	RATE
20	Kaiser	Valerie	624 Randall	Grove	FL	33321	$20,542.50	1	0.05
35	Hull	Richard	532 Jackson	Sheldon	FL	33553	$39,216.00	2	0.07
65	Perez	Juan	1626 Taylor	Fillmore	FL	33336	$23,487.00	1	0.05

FIGURE 2-4 REP table with a PAY_CLASS column

Suppose one of the policies at Premiere Products is that all sales reps in any given pay class earn their commissions at the same rate. To describe this situation, you could say that a sales rep's pay class *determines* his or her commission rate. Alternatively, you could say that a sales rep's commission rate *depends on* his or her pay class. This phrasing uses the words *determines* and *depends on* in the same way that you describe functional dependency. If you wanted to be formal, you would precede either expression with the word *functionally*. For example, you might say, "A sales rep's pay class *functionally determines* his or her commission rate," and "A sales rep's commission rate *functionally depends on* his or her pay class." You can also define functional dependency by saying that when you know a sales rep's pay class, you can determine his or her commission rate.

In a relational database, column B is **functionally dependent** on another column (or a collection of columns), A, if at any point in time a value for A determines a single value for B. You can think of this as follows: when you are given a value for A, do you know that you can find a single value for B? If so, B is functionally dependent on A (often written as A → B). If B is functionally dependent on A, you also can say that A **functionally determines** B.

At Premiere Products, is the LAST_NAME column in the REP table functionally dependent on the REP_NUM column? Yes, it is. If you are given a value for REP_NUM, such as 20, there is a *single* LAST_NAME, Kaiser, associated with it. This is represented as:

```
REP_NUM → LAST_NAME
```

Q & A

Question: In the CUSTOMER table, is CUSTOMER_NAME functionally dependent on REP_NUM?
Answer: No. Given the REP_NUM 20, for example, you would not be able to find a single customer name, because 20 appears on more than one row in the table.

Q & A

Question: In the ORDER_LINE table, is NUM_ORDERED functionally dependent on ORDER_NUM?
Answer: No. An ORDER_NUM might be associated with several items in an order, so having just an ORDER_NUM does not provide enough information.

Q & A

Question: Is NUM_ORDERED functionally dependent on PART_NUM?
Answer: No. Again, just as with ORDER_NUM, a PART_NUM might be associated with several items in an order, so PART_NUM does not provide enough information.

Q & A

Question: On which columns in the ORDER_LINE table is NUM_ORDERED functionally dependent?
Answer: To determine a value for NUM_ORDERED, you need both an order number and a part number. In other words, NUM_ORDERED is functionally dependent on the combination (formally called the **concatenation**) of ORDER_NUM and PART_NUM. That is, given an order number *and* a part number, you can find a single value for NUM_ORDERED.

At this point, a question naturally arises: how do you determine functional dependencies? Can you determine them by looking at sample data, for example? The answer is no.

Consider the REP table in Figure 2-5, in which last names are unique. It is very tempting to say that LAST_NAME functionally determines STREET, CITY, STATE, and ZIP (or equivalently that STREET, CITY, STATE, and ZIP are all functionally dependent on LAST_NAME). After all, given the last name of a rep, you can find the single address.

REP

REP_ NUM	LAST_ NAME	FIRST_ NAME	STREET	CITY	STATE	ZIP	COMMISSION	RATE
20	Kaiser	Valerie	624 Randall	Grove	FL	33321	$20,542.50	0.05
35	Hull	Richard	532 Jackson	Sheldon	FL	33553	$39,216.00	0.07
65	Perez	Juan	1626 Taylor	Fillmore	FL	33336	$23,487.00	0.05

FIGURE 2-5 REP table

What happens when rep 85, whose last name is also Kaiser, is added to the database? You then have the situation illustrated in Figure 2-6. Because there are now two reps with the last name of Kaiser, you can no longer find a single address using a rep's last name—you were misled by the original data. The only way to determine functional dependencies is to examine the user's policies. This process can involve discussions with users, an examination of user documentation, and so on. For example, if managers at Premiere Products have a policy never to hire two reps with the same last name, then LAST_NAME would indeed determine the other columns. Without such a policy, however, LAST_NAME would not determine the other columns.

REP

REP_ NUM	LAST_ NAME	FIRST_ NAME	STREET	CITY	STATE	ZIP	COMMISSION	RATE
20	Kaiser	Valerie	624 Randall	Grove	FL	33321	$20,542.50	0.05
35	Hull	Richard	532 Jackson	Sheldon	FL	33553	$39,216.00	0.07
65	Perez	Juan	1626 Taylor	Fillmore	FL	33336	$23,487.00	0.05
85	Kaiser	William	172 Bahia	Norton	FL	39281	$0.00	0.05

FIGURE 2-6 REP table with two reps named Kaiser

PRIMARY KEYS

Another important database design concept is the primary key. In the simplest terms, the **primary key** is the unique identifier for a table. For example, the REP_NUM column is the unique identifier for the REP table. Given a rep number in the table, such as 20, there will only be one row on which that rep number occurs. Thus, the rep number 20 uniquely identifies a row (in this case, the first row, and the rep named Valerie Kaiser).

In this text, the definition of primary key needs to be more precise than a unique identifier for a table. Specifically, column A (or a collection of columns) is the primary key for a table if:

Property 1. *All* columns in the table are functionally dependent on A.
Property 2. No subcollection of the columns in A (assuming A is a collection of columns and not just a single column) also has property 1.

Q & A

Question: Is the CLASS column the primary key for the PART table?
Answer: No, because the other columns are not functionally dependent on CLASS. Given the class HW, for example, you cannot determine a part number, description, or anything else, because there are several rows on which the class is HW.

Q & A

Question: Is the CUSTOMER_NUM column the primary key for the CUSTOMER table?
Answer: Yes, because Premiere Products assigns unique customer numbers. A specific customer number cannot appear on more than one row. Thus, all columns in the CUSTOMER table are functionally dependent on CUSTOMER_NUM.

Q & A

Question: Is the ORDER_NUM column the primary key for the ORDER_LINE table?
Answer: No, because it does not functionally determine either NUM_ORDERED or QUOTED_PRICE.

Q & A

Question: Is the combination of the ORDER_NUM and PART_NUM columns the primary key for the ORDER_LINE table?
Answer: Yes, because you can determine all columns by this combination of columns, and, further, neither the ORDER_NUM nor the PART_NUM alone has this property.

Q & A

Question: Is the combination of the PART_NUM and DESCRIPTION columns the primary key for the PART table?
Answer: No. Although it is true that you can determine all columns in the PART table by this combination, PART_NUM alone also has this property.

You can indicate a table's primary key with a shorthand representation of a database by underlining the column or collection of columns that comprise the primary key. The complete shorthand representation for the Premiere Products database is:

```
REP (REP_NUM, LAST_NAME, FIRST_NAME, STREET,
     CITY, STATE, ZIP, COMMISSION, RATE)
CUSTOMER (CUSTOMER_NUM, CUSTOMER_NAME, STREET,
     CITY, STATE, ZIP, BALANCE, CREDIT_LIMIT,
     REP_NUM)
ORDERS (ORDER_NUM, ORDER_DATE, CUSTOMER_NUM)
ORDER_LINE (ORDER_NUM, PART_NUM, NUM_ORDERED,
     QUOTED_PRICE)
PART (PART_NUM, DESCRIPTION, ON_HAND, CLASS,
     WAREHOUSE, PRICE)
```

NOTE

Sometimes you might identify one or more columns that you can use as a table's primary key. For example, if the Premiere Products database also included an EMPLOYEE table that contains employee numbers and Social Security numbers, either the employee number or the Social Security number could serve as the table's primary key. In this case, both columns are referred to as candidate keys. Like a primary key, a **candidate key** is a column or collection of columns on which all columns in the table are functionally dependent—the definition for primary key really defines candidate key as well. From all the candidate keys, you would choose one to be the primary key.

NOTE

According to the definition of a candidate key, a Social Security number is a legitimate primary key. Many databases, such as those that store data about students at a college or university or those that store data about employees at a company, store a person's Social Security number as a primary key. However, many institutions and organizations are moving away from using Social Security numbers as primary keys because of privacy issues. Instead of using Social Security numbers, many institutions and organizations use unique student numbers or employee numbers as primary keys.

NOTE

Some institutions prefer to assign values to use as primary keys for items such as customer numbers, part numbers, and student numbers. Others simply let the computer generate the values. In this case, the DBMS simply assigns the next available value. For example, if the database has already assigned customer numbers 1000 through 1436, it assigns the next new customer added to the database the customer number 1437.

DATABASE DESIGN

This section presents a specific method you can follow to design a database when given a set of requirements that the database must support. The determination of the requirements is part of the process known as systems analysis. A systems analyst interviews users, examines existing and proposed documents, and examines organizational policies to determine exactly the type of data needs the database must support. This text does not cover this analysis. Rather, it focuses on how to take the set of requirements that this process produces and determine the appropriate database design.

After presenting the database design method, this section presents a sample set of requirements and illustrates the design method by designing a database to satisfy these requirements.

Design Method

To design a database for a set of requirements, complete the following steps:

1. Read the requirements, identify the entities (objects) involved, and name the entities. For example, when the design involves departments and employees, you might use the entity names DEPARTMENT and EMPLOYEE. When the design involves customers and sales reps, you might use the entity names CUSTOMER and REP.
2. Identify the unique identifiers for the entities you identified in Step 1. For example, when one of the entities is PART, determine what information is required to uniquely identify each individual part. In other words, what information does the organization use to distinguish one part from another? For a PART entity, the unique identifier for each part might be a PART_NUM; for a CUSTOMER entity, the unique identifier might be a CUSTOMER_NUM. When no unique identifier is available from the data you know about the entity, you need to create one. For example, you might use a unique number to identify parts when no part numbers exist.
3. Identify the attributes for all the entities. These attributes become the columns in the tables. It is possible for two or more entities to contain the same attributes. At Premiere Products, for example, reps and customers both have addresses, cities, states, and zip codes. To clarify this duplication of attributes, follow the name of the attribute with the corresponding entity in parentheses. Thus, ADDRESS (CUSTOMER) is a customer address and ADDRESS (REP) is a sales rep address.
4. Identify the functional dependencies that exist among the attributes. Ask yourself the following question: if you know a unique value for an attribute, do you also

know the unique values for other attributes? For example, when you have the three attributes REP_NUM, LAST_NAME, and FIRST_NAME and you know a unique value for REP_NUM, do you also know a unique value for LAST_NAME and FIRST_NAME? If so, then LAST_NAME and FIRST_NAME are functionally dependent on REP_NUM (REP_NUM → LAST_NAME, FIRST_NAME).

5. Use the functional dependencies to identify the tables by placing each attribute with the attribute or minimum combination of attributes on which it is functionally dependent. The attribute or attributes for an entity on which all other attributes are dependent will be the primary key of the table. The remaining attributes will be the other columns in the table. Once you have determined all the columns in the table, you can give the table an appropriate name. Usually the name will be the same as the name you identified for the entity in Step 1.
6. Identify any relationships between tables. In some cases, you might be able to determine the relationships directly from the requirements. It might be clear, for example, that one rep is related to many customers and that each customer is related to exactly one rep. When it is not, look for matching columns in the tables you created. For example, if both the REP table and the CUSTOMER table contain a REP_NUM column and the values in these columns must match, you know that reps and customers are related. The fact that the REP_NUM column is the primary key in the REP table tells you that the REP table is the "one" part of the relationship and the CUSTOMER table is the "many" part of the relationship.

In the next section, you will apply this process to produce the design for the Premiere Products database using the collection of requirements that this database must support.

Database Design Requirements

The analyst has interviewed users and examined documents at Premiere Products and has determined that the database must support the following requirements:

1. For a sales rep, store the sales rep's number, last name, first name, street address, city, state, zip code, total commission, and commission rate.
2. For a customer, store the customer's number, name, street address, city, state, zip code, balance, and credit limit. In addition, store the number, last name, and first name of the sales rep who represents this customer. The analyst has also determined that a sales rep can represent many customers, but a customer must have exactly one sales rep (in other words, a sales rep must represent a customer; a customer cannot be represented by zero or more than one sales reps).
3. For a part, store the part's number, description, units on hand, item class, the number of the warehouse in which the part is located, and the price. All units of a particular part are stored in the same warehouse.
4. For an order, store the order number, order date, the number and name of the customer that placed the order, and the number of the sales rep who represents that customer.

5. For each line item within an order, store the part number and description, the number ordered, and the quoted price. The analyst also obtained the following information concerning orders:
   a. There is only one customer per order.
   b. On a given order, there is at most one line item for a given part. For example, part DR93 cannot appear on several lines within the same order.
   c. The quoted price might differ from the actual price when the sales rep discounts a certain part on a specific order.

Database Design Process Example

The following steps apply the design process to the requirements for Premiere Products to produce the appropriate database design:

Step 1: There appear to be four entities: reps, customers, parts, and orders. The names assigned to these entities are REP, CUSTOMER, PART, and ORDERS, respectively.

Step 2: From the collection of entities, review the data and determine the unique identifier for each entity. For the REP, CUSTOMER, PART, and ORDERS entities, the unique identifiers are the rep number, customer number, part number, and order number, respectively. These unique identifiers are named REP_NUM, CUSTOMER_NUM, PART_NUM, and ORDER_NUM, respectively.

Step 3: The attributes mentioned in the first requirement all refer to sales reps. The specific attributes mentioned in the requirement are the sales rep's number, name, street address, city, state, zip code, total commission, and commission rate. Assigning appropriate names to these attributes produces the following list:

```
REP_NUM
LAST_NAME
FIRST_NAME
STREET
CITY
STATE
ZIP
COMMISSION
RATE
```

The attributes mentioned in the second requirement refer to customers. The specific attributes are the customer's number, name, street address, city, state, zip code, balance, and credit limit. The requirement also mentions the number, first name, and last name of the sales rep who represents this customer. Assigning appropriate names to these attributes produces the following list:

```
CUSTOMER_NUM
CUSTOMER_NAME
STREET
CITY
STATE
ZIP
BALANCE
CREDIT_LIMIT
REP_NUM
LAST_NAME
FIRST_NAME
```

There are attributes named STREET, CITY, STATE, and ZIP for sales reps as well as attributes named STREET, CITY, STATE, and ZIP for customers. To distinguish these attributes in the final collection, follow the name of the attribute by the name of the corresponding entity. For example, the street for a sales rep is STREET (REP) and the street for a customer is STREET (CUSTOMER).

The attributes mentioned in the third requirement refer to parts. The specific attributes are the part's number, description, units on hand, item class, the number of the warehouse in which the part is located, and the price. Assigning appropriate names to these attributes produces the following list:

```
PART_NUM
DESCRIPTION
ON_HAND
CLASS
WAREHOUSE
PRICE
```

The attributes mentioned in the fourth requirement refer to orders. The specific attributes include the order number, order date, number and name of the customer that placed the order, and number of the sales rep who represents the customer. Assigning appropriate names to these attributes produces the following list:

```
ORDER_NUM
ORDER_DATE
CUSTOMER_NUM
CUSTOMER_NAME
REP_NUM
```

The specific attributes associated with the statement in the requirements concerning line items are the order number (to determine the order to which the line item corresponds), part number, description, number ordered, and quoted price. If the quoted price must be the same as the price, you could simply call it PRICE. According to requirement 5c, however, the quoted price might differ from the price, so you must add the quoted price to the list. Assigning appropriate names to these attributes produces the following list:

```
ORDER_NUM
PART_NUM
DESCRIPTION
NUM_ORDERED
QUOTED_PRICE
```

The complete list grouped by entity is as follows:

REP

```
REP_NUM
LAST_NAME
FIRST_NAME
STREET (REP)
CITY (REP)
STATE (REP)
ZIP (REP)
COMMISSION
RATE
```

CUSTOMER
CUSTOMER_NUM
CUSTOMER_NAME
STREET (CUSTOMER)
CITY (CUSTOMER)
STATE (CUSTOMER)
ZIP (CUSTOMER)
BALANCE
CREDIT_LIMIT
REP_NUM
LAST_NAME
FIRST_NAME

PART
PART_NUM
DESCRIPTION
ON_HAND
CLASS
WAREHOUSE
PRICE

ORDER
ORDER_NUM
ORDER_DATE
CUSTOMER_NUM
CUSTOMER_NAME
REP_NUM

For line items within an order
ORDER_NUM
PART_NUM
DESCRIPTION
NUM_ORDERED
QUOTED_PRICE

Step 4: The fact that the unique identifier for sales reps is the rep number gives the following functional dependencies:

```
REP_NUM → LAST_NAME, FIRST_NAME, STREET (REP), CITY (REP),
       STATE (REP), ZIP (REP), COMMISSION, RATE
```

This notation indicates that the LAST_NAME, FIRST_NAME, STREET (REP), CITY (REP), STATE (REP), ZIP (REP), COMMISSION, and RATE are all functionally dependent on REP_NUM.

The fact that the unique identifier for customers is the customer number gives the following functional dependencies:

```
CUSTOMER_NUM → CUSTOMER_NAME, STREET (CUSTOMER),
      CITY (CUSTOMER), STATE (CUSTOMER), ZIP (CUSTOMER),
      BALANCE, CREDIT_LIMIT, REP_NUM, LAST_NAME, FIRST_NAME
```

Q & A

Question: Do you really need to include the last name and first name of a sales rep in the list of attributes determined by the customer number?
Answer: There is no need to include them in this list, because they both can be determined from the sales rep number and are already included in the list of attributes determined by REP_NUM.

Thus, the functional dependencies for the CUSTOMER entity are as follows:

```
CUSTOMER_NUM → CUSTOMER_NAME, STREET (CUSTOMER),
     CITY (CUSTOMER), STATE (CUSTOMER), ZIP (CUSTOMER),
     BALANCE, CREDIT_LIMIT, REP_NUM
```

The fact that the unique identifier for parts is the part number gives the following functional dependencies:

```
PART_NUM → DESCRIPTION, ON_HAND, CLASS, WAREHOUSE, PRICE
```

The fact that the unique identifier for orders is the order number gives the following functional dependencies:

```
ORDER_NUM → ORDER_DATE, CUSTOMER_NUM, CUSTOMER_NAME,
     REP_NUM
```

Q & A

Question: Do you really need to include the name of a customer and the number of the customer's rep in the list of attributes determined by the order number?
Answer: There is no need to include the customer name and the rep number in this list, because you can determine them from the customer number and they are already included in the list of attributes determined by CUSTOMER_NUM.

The functional dependencies for the ORDERS entity are as follows:

```
ORDER_NUM → ORDER_DATE, CUSTOMER_NUM
```

The final attributes to be examined are those associated with the line items within the order: PART_NUM, DESCRIPTION, NUM_ORDERED, and QUOTED_PRICE.

Q & A

Question: Why aren't NUM_ORDERED and QUOTED_PRICE included in the list of attributes determined by the order number?
Answer: To uniquely identify a particular value for NUM_ORDERED or QUOTED_PRICE, ORDER_NUM alone is not sufficient. It requires the combination of ORDER_NUM and PART_NUM.

The following shorthand representation indicates that the combination of ORDER_NUM and PART_NUM functionally determines NUM_ORDERED and QUOTED_PRICE:

```
ORDER_NUM, PART_NUM → NUM_ORDERED, QUOTED_PRICE
```

Q & A

Question: Does DESCRIPTION need to be included in this list?
Answer: No, because DESCRIPTION can be determined by the PART_NUMBER alone, and it already appears in the list of attributes dependent on the PART_NUM.

The complete list of functional dependencies is as follows:

```
REP_NUM → LAST_NAME, FIRST_NAME, STREET (REP), CITY (REP),
     STATE (REP), ZIP(REP), COMMISSION, RATE
CUSTOMER_NUM → CUSTOMER_NAME, STREET (CUSTOMER),
     CITY (CUSTOMER), STATE (CUSTOMER), ZIP (CUSTOMER),
     BALANCE, CREDIT_LIMIT, REP_NUM
PART_NUM → DESCRIPTION, ON_HAND, CLASS, WAREHOUSE, PRICE
ORDER_NUM → ORDER_DATE, CUSTOMER_NUM
ORDER_NUM, PART_NUM → NUM_ORDERED, QUOTED_PRICE
```

Step 5: Using the functional dependencies, you can create tables with the attribute(s) to the left of the arrow being the primary key and the items to the right of the arrow being the other columns. For relations corresponding to those entities identified in Step 1, you can use the name you already determined. Because you did not identify any entity that had a unique identifier that was the combination of ORDER_NUM and PART_NUM, you need to assign a name to the table whose primary key consists of these two columns. Because this table represents the individual lines within an order, the name ORDER_LINE is a good choice. The final collection of tables is as follows:

```
REP (REP_NUM, LAST_NAME, FIRST_NAME, STREET,
     CITY, STATE, ZIP, COMMISSION, RATE)
CUSTOMER (CUSTOMER_NUM, CUSTOMER_NAME, STREET,
     CITY, STATE, ZIP, BALANCE, CREDIT_LIMIT,
     REP_NUM)
PART (PART_NUM, DESCRIPTION, ON_HAND, CLASS,
      WAREHOUSE, PRICE)
ORDERS (ORDER_NUM, ORDER_DATE, CUSTOMER_NUM)
ORDER_LINE (ORDER_NUM, PART_NUM, NUM_ORDERED,
      QUOTED_PRICE)
```

Step 6: Examining the tables and identifying common columns gives the following list of relationships between the tables:

- The CUSTOMER and REP tables are related using the REP_NUM columns. Because the REP_NUM column is the primary key for the REP table, this indicates a one-to-many relationship between REP and CUSTOMER (one rep to many customers).
- The ORDERS and CUSTOMER tables are related using the CUSTOMER_NUM columns. Because the CUSTOMER_NUM column is the primary key for the CUSTOMER table, this indicates a one-to-many relationship between CUSTOMER and ORDERS (one customer to many orders).
- The ORDER_LINE and ORDERS tables are related using the ORDER_NUM columns. Because the ORDER_NUM column is the primary key for the ORDERS table, this indicates a one-to-many relationship between ORDERS and ORDER_LINE (one order to many order lines).
- The ORDER_LINE and PART tables are related using the PART_NUM columns. Because the PART_NUM column is the primary key for the PART table, this indicates a one-to-many relationship between PART and ORDER_LINE (one part to many order lines).

NORMALIZATION

After creating the database design, you must analyze it to make sure it is free of potential problems. To do so, you follow a process called **normalization**, in which you identify the existence of potential problems, such as data duplication and redundancy, and implement ways to correct these problems.

The goal of normalization is to convert **unnormalized relations** (tables that satisfy the definition of a relation except that they might contain repeating groups) into various types of **normal forms**. A table in a particular normal form possesses a certain desirable collection of properties. Although there are several normal forms, the most common are first normal form, second normal form, and third normal form. Normalization is a process in which a table that is in first normal form is better than a table that is not in first normal form, a table that is in second normal form is better than one that is in first normal form, and so on. The goal of this process is to allow you to take a table or collection of tables and produce a new collection of tables that represents the same information but is free of problems.

First Normal Form

According to the definition of a relation, a relation (table) cannot contain a repeating group in which multiple entries exist on a single row. However, in the database design process, you might create a table that has all the other properties of a relation, but contains a repeating group. Removing repeating groups is the starting point when converting an unnormalized collection of data into a table that is in first normal form. A table (relation) is in **first normal form (1NF)** when it does not contain a repeating group.

For example, in the design process you might create the following ORDERS table, in which there is a repeating group consisting of PART_NUM and NUM_ORDERED. The notation for this table is as follows:

```
ORDERS (ORDER_NUM, ORDER_DATE, (PART_NUM, NUM_ORDERED) )
```

This notation describes a table named ORDERS that consists of a primary key, ORDER_NUM, and a column named ORDER_DATE. The inner parentheses indicate a repeating group that contains two columns, PART_NUM and NUM_ORDERED. This table contains one row per order with values in the PART_NUM and NUM_ORDERED columns for each order with the number ORDER_NUM and placed on ORDER_DATE. Figure 2-7 shows a single order with multiple combinations of a part number and a corresponding number of units ordered.

ORDERS

ORDER_ NUM	ORDER_ DATE	PART_ NUM	NUM_ ORDERED
21608	10/20/2010	AT94	11
21610	10/20/2010	DR93 DW11	1 1
21613	10/21/2010	KL62	4
21614	10/21/2010	KT03	2
21617	10/23/2010	BV06 CD52	2 4
21619	10/23/2010	DR93	1
21623	10/23/2010	KV29	2

FIGURE 2-7 Unnormalized order data

To convert the table to first normal form, you remove the repeating group as follows:

```
ORDERS (ORDER_NUM, ORDER_DATE, PART_NUM, NUM_ORDERED)
```

Figure 2-8 shows the table in first normal form.

ORDERS

ORDER_ NUM	ORDER_ DATE	PART_ NUM	NUM_ ORDERED
21608	10/20/2010	AT94	11
21610	10/20/2010	DR93	1
21610	10/20/2010	DW11	1
21613	10/21/2010	KL62	4
21614	10/21/2010	KT03	2
21617	10/23/2010	BV06	2
21617	10/23/2010	CD52	4
21619	10/23/2010	DR93	1
21623	10/23/2010	KV29	2

FIGURE 2-8 Order data converted to first normal form

In Figure 2-7, the second row indicates that part DR93 and part DW11 are both included in order 21610. In Figure 2-8, this information is represented by *two* rows, the second and third. The primary key for the unnormalized ORDERS table was the ORDER_NUM column alone. The primary key for the normalized table is now the combination of the ORDER_NUM and PART_NUM columns.

When you convert an unnormalized table to a table in first normal form, the primary key of the table in first normal form is usually the primary key of the unnormalized table concatenated with the key for the repeating group, which is the column in the repeating group that distinguishes one occurrence of the repeating group from another within a given row in the table. In the ORDERS table, PART_NUM was the key to the repeating group and ORDER_NUM was the primary key for the table. When converting the unnormalized data to first normal form, the primary key becomes the concatenation of the ORDER_NUM and PART_NUM columns.

Second Normal Form

The following ORDERS table is in first normal form, because it does not contain a repeating group:

```
ORDERS (ORDER_NUM, ORDER_DATE, PART_NUM, DESCRIPTION,
     NUM_ORDERED, QUOTED_PRICE)
```

The table contains the following functional dependencies:

```
ORDER_NUM → ORDER_DATE
PART_NUM → DESCRIPTION
ORDER_NUM, PART_NUM → NUM_ORDERED, QUOTED_PRICE
```

This notation indicates that ORDER_NUM alone determines ORDER_DATE, and PART_NUM alone determines DESCRIPTION, but it requires *both* an ORDER_NUM *and* a PART_NUM to determine either NUM_ORDERED or QUOTED_PRICE. Consider the sample of this table shown in Figure 2-9.

ORDERS

ORDER_ NUM	ORDER_ DATE	PART_ NUM	DESCRIPTION	NUM_ ORDERED	QUOTED_ PRICE
21608	10/20/2010	AT94	Iron	11	$21.95
21610	10/20/2010	DR93	Gas Range	1	$495.00
21610	10/20/2010	DW11	Washer	1	$399.99
21613	10/21/2010	KL62	Dryer	4	$329.95
21614	10/21/2010	KT03	Dishwasher	2	$595.00
21617	10/23/2010	BV06	Home Gym	2	$12.95
21617	10/23/2010	CD52	Microwave Oven	4	$150.00
21619	10/23/2010	DR93	Gas Range	1	$495.00
21623	10/23/2010	KV29	Treadmill	2	$325.99

FIGURE 2-9 Sample ORDERS table

Although the ORDERS table is in first normal form (because it contains no repeating groups), problems exist within the table that require you to restructure it.

The description of a specific part, DR93 for example, occurs twice in the table. This duplication (formally called **redundancy**) causes several problems. It is certainly wasteful of space, but that is not nearly as serious as some of the other problems. These other problems are called **update anomalies** and they fall into four categories:

1. **Updates:** If you need to change to the description of part DR93, you must change it twice—once in each row on which part DR93 appears. Updating the part description more than once makes the update process much more cumbersome and time consuming.
2. **Inconsistent data:** There is nothing about the design that prohibits part DR93 from having two *different* descriptions in the database. In fact, if part DR93 occurs on 20 rows in the table, it is possible for this part to have 20 different descriptions in the database.
3. **Additions:** When you try to add a new part and its description to the database, you will face a real problem. Because the primary key for the ORDERS table consists of both an ORDER_NUM and a PART_NUM, you need values for both of these columns to add a new row to the table. If you add a part to the table that does not yet have any orders, what do you use for an ORDER_NUM? The only solution is to create a dummy ORDER_NUM and then replace

it with a real ORDER_NUM once an order for this part is actually received. Certainly this is not an acceptable solution.

4. **Deletions:** If you delete order 21608 from the database and it is the only order that contains part AT94, deleting the order also deletes all information about part AT94. For example, you would no longer know that part AT94 is an iron.

These problems occur because you have a column, DESCRIPTION, that is dependent on only a portion of the primary key, PART_NUM, and *not* on the complete primary key. This situation leads to the definition of second normal form. Second normal form represents an improvement over first normal form because it eliminates update anomalies in these situations. A table (relation) is in **second normal form (2NF)** when it is in first normal form and no **nonkey column** (that is, a column that is not part of the primary key) is dependent on only a portion of the primary key.

NOTE

When the primary key of a table contains only a single column, the table is automatically in second normal form.

You can identify the fundamental problem with the ORDERS table: it is not in second normal form. Although it is important to identify the problem, what you really need is a method to *correct* it; you want to be able to convert tables to second normal form. First, take each subset of the set of columns that make up the primary key, and begin a new table with this subset as its primary key. For the ORDERS table, the new design is:

```
(ORDER_NUM,
(PART_NUM,
(ORDER_NUM, PART_NUM,
```

Next, place each of the other columns with the appropriate primary key; that is, place each one with the minimal collection of columns on which it depends. For the ORDERS table, add the new columns as follows:

```
(ORDER_NUM, ORDER_DATE)
(PART_NUM, DESCRIPTION)
(ORDER_NUM, PART_NUM, NUM_ORDERED, QUOTED_PRICE)
```

Each of these new tables is given a descriptive name based on the meaning and contents of the table, such as ORDERS, PART, and ORDER_LINE. Figure 2-10 shows samples of these tables.

ORDERS

ORDER_ NUM	ORDER_ DATE	PART_ NUM	DESCRIPTION	NUM_ ORDERED	QUOTED_ PRICE
21608	10/20/2010	AT94	Iron	11	$21.95
21610	10/20/2010	DR93	Gas Range	1	$495.00
21610	10/20/2010	DW11	Washer	1	$399.99
21613	10/21/2010	KL62	Dryer	4	$329.95
21614	10/21/2010	KT03	Dishwasher	2	$595.00
21617	10/23/2010	BV06	Home Gym	2	$12.95
21617	10/23/2010	CD52	Microwave Oven	4	$150.00
21619	10/23/2010	DR93	Gas Range	1	$495.00
21623	10/23/2010	KV29	Treadmill	2	$325.99

ORDERS

ORDER_ NUM	ORDER_ DATE
21608	10/20/2010
21610	10/20/2010
21613	10/21/2010
21614	10/21/2010
21617	10/23/2010
21619	10/23/2010
21623	10/23/2010

PART

PART_ NUM	DESCRIPTION
AT94	Iron
BV06	Home Gym
CD52	Microwave Oven
DL71	Cordless Drill
DR93	Gas Range
DW11	Washer
FD21	Stand Mixer
KL62	Dryer
KT03	Dishwasher
KV29	Treadmill

ORDER_LINE

ORDER_ NUM	PART_ NUM	NUM_ ORDERED	QUOTED_ PRICE
21608	AT94	11	$21.95
21610	DR93	1	$495.00
21610	DW11	1	$399.99
21613	KL62	4	$329.95
21614	KT03	2	$595.00
21617	BV06	2	$12.95
21617	CD52	4	$150.00
21619	DR93	1	$495.00
21623	KV29	2	$325.99

FIGURE 2-10 ORDERS table converted to second normal form

In Figure 2-10, converting the original ORDERS table to a new ORDERS table, a PART table, and an ORDER_LINE table eliminates the update anomalies. A description appears only once for each part, so you do not have the redundancy that existed in the original table design. Changing the description of part DR93 from Gas Range to Deluxe Range, for example, is now a simple process involving a single change. Because the description for a part occurs in a single place, it is not possible to have multiple descriptions for a single part in the database at the same time.

To add a new part and its description, you create a new row in the PART table, regardless of whether that part has pending or actual orders. Also, deleting order 21608 does not delete part number AT94 from the database because it still exists in the PART table. Finally, you have not lost any information by converting the ORDERS table to second normal form. You can reconstruct the data in the original table from the data in the new tables.

Third Normal Form

Problems can still exist with tables that are in second normal form. For example, suppose that you create the following CUSTOMER table:

```
CUSTOMER (CUSTOMER_NUM, CUSTOMER_NAME, BALANCE, CREDIT_LIMIT,
     REP_NUM, LAST_NAME, FIRST_NAME)
```

This table has the following functional dependencies:

```
CUSTOMER_NUM → CUSTOMER_NAME, BALANCE, CREDIT_LIMIT,
     REP_NUM, LAST_NAME, FIRST_NAME
REP_NUM → LAST_NAME, FIRST_NAME
```

CUSTOMER_NUM determines all the other columns. In addition, REP_NUM determines LAST_NAME and FIRST_NAME.

When a table's primary key is a single column, the table is automatically in second normal form. (If the table were not in second normal form, some column would be dependent on only a *portion* of the primary key, which is impossible when the primary key is just one column.) Thus, the CUSTOMER table is in second normal form.

Although this table is in second normal form, Figure 2-11 shows that it still possesses update problems similar to those identified for the ORDERS table shown in Figure 2-9. In Figure 2-11, the sales rep name occurs many times in the table.

CUSTOMER

CUSTOMER_ NUM	CUSTOMER_NAME	BALANCE	CREDIT_ LIMIT	REP_ NUM	LAST_ NAME	FIRST_ NAME
148	Al's Appliance and Sport	$6,550.00	$7,500.00	20	Kaiser	Valerie
282	Brookings Direct	$431.50	$10,000.00	35	Hull	Richard
356	Ferguson's	$5,785.00	$7,500.00	65	Perez	Juan
408	The Everything Shop	$5,285.25	$5,000.00	35	Hull	Richard
462	Bargains Galore	$3,412.00	$10,000.00	65	Perez	Juan
524	Kline's	$12,762.00	$15,000.00	20	Kaiser	Valerie
608	Johnson's Department Store	$2,106.00	$10,000.00	65	Perez	Juan
687	Lee's Sport and Appliance	$2,851.00	$5,000.00	35	Hull	Richard
725	Deerfield's Four Seasons	$248.00	$7,500.00	35	Hull	Richard
842	All Season	$8,221.00	$7,500.00	20	Kaiser	Valerie

FIGURE 2-11 Sample CUSTOMER table

The redundancy of including a sales rep number and name in the CUSTOMER table results in the same set of problems that existed for the ORDERS table. In addition to the problem of wasted space, you have the following update anomalies:

1. **Updates:** Changing the sales rep name requires changes to multiple rows in the table.
2. **Inconsistent data:** The design does not prohibit multiple iterations of sales rep names in the database. For example, a sales rep might represent 20 customers and his name might be entered 20 different ways in the table.
3. **Additions:** To add sales rep 87 (Emily Daniels) to the database, she must represent at least one customer. If Emily does not yet represent any customers, you either cannot record the fact that her name is Emily Daniels or you must create a fictitious customer for her to represent until she represents an actual customer. Neither of these solutions is desirable.
4. **Deletions:** If you delete all the customers of sales rep 35 from the database, you will also lose all information about sales rep 35.

These update anomalies are due to the fact that REP_NUM determines LAST_NAME and FIRST_NAME, but REP_NUM is not the primary key. As a result, the same REP_NUM and consequently the same LAST_NAME and FIRST_NAME can appear on many different rows.

You have seen that tables in second normal form represent an improvement over tables in first normal form, but to eliminate problems with tables in second normal form, you need an even better strategy for creating tables. Third normal form provides that strategy.

Before looking at third normal form, however, you need to become familiar with the special name that is given to any column that determines another column (like REP_NUM in the CUSTOMER table). Any column (or collection of columns) that determines another column is called a **determinant**. A table's primary key is a determinant. In fact, by definition, any candidate key is a determinant. (Remember that a candidate key is a column or collection of columns that could function as the primary key.) In Figure 2-11, REP_NUM is a determinant, but it is not a candidate key, and that is the problem.

A table is in **third normal form (3NF)** when it is in second normal form and the only determinants it contains are candidate keys.

NOTE

This text's definition of third normal form is not the original definition. This more recent definition, which is preferable to the original, is often referred to as **Boyce-Codd normal form (BCNF)** when it is important to make a distinction between this definition and the original definition. This text does not make such a distinction but will take this to be the definition of third normal form.

Now you have identified the problem with the CUSTOMER table: it is not in third normal form. There are several steps for converting tables to third normal form.

First, for each determinant that is not a candidate key, remove from the table the columns that depend on this determinant (but do not remove the determinant). Next, create a new table containing all the columns from the original table that depend on this determinant. Finally, make the determinant the primary key of this new table.

In the CUSTOMER table, for example, remove LAST_NAME and FIRST_NAME because they depend on the determinant REP_NUM, which is not a candidate key. A new table is formed, consisting of REP_NUM as the primary key, and the columns LAST_NAME and FIRST_NAME, as follows:

```
CUSTOMER (CUSTOMER_NUM, CUSTOMER_NAME, BALANCE,
     CREDIT_LIMIT, REP_NUM)
```

and

```
REP (REP_NUM, LAST_NAME, FIRST_NAME)
```

Figure 2-12 shows the original CUSTOMER table and the tables created when converting the original table to third normal form.

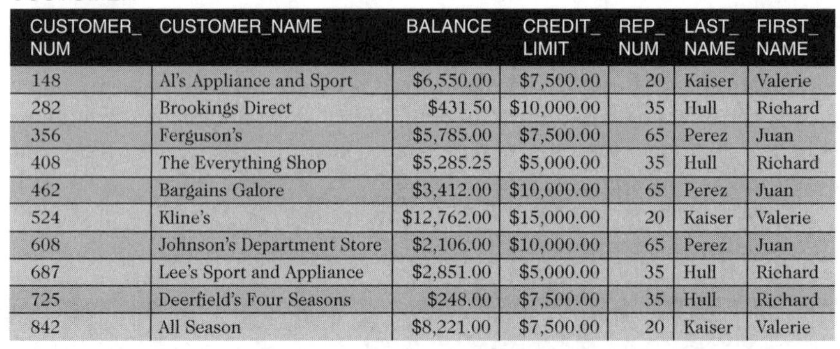

CUSTOMER

CUSTOMER_NUM	CUSTOMER_NAME	BALANCE	CREDIT_LIMIT	REP_NUM	LAST_NAME	FIRST_NAME
148	Al's Appliance and Sport	$6,550.00	$7,500.00	20	Kaiser	Valerie
282	Brookings Direct	$431.50	$10,000.00	35	Hull	Richard
356	Ferguson's	$5,785.00	$7,500.00	65	Perez	Juan
408	The Everything Shop	$5,285.25	$5,000.00	35	Hull	Richard
462	Bargains Galore	$3,412.00	$10,000.00	65	Perez	Juan
524	Kline's	$12,762.00	$15,000.00	20	Kaiser	Valerie
608	Johnson's Department Store	$2,106.00	$10,000.00	65	Perez	Juan
687	Lee's Sport and Appliance	$2,851.00	$5,000.00	35	Hull	Richard
725	Deerfield's Four Seasons	$248.00	$7,500.00	35	Hull	Richard
842	All Season	$8,221.00	$7,500.00	20	Kaiser	Valerie

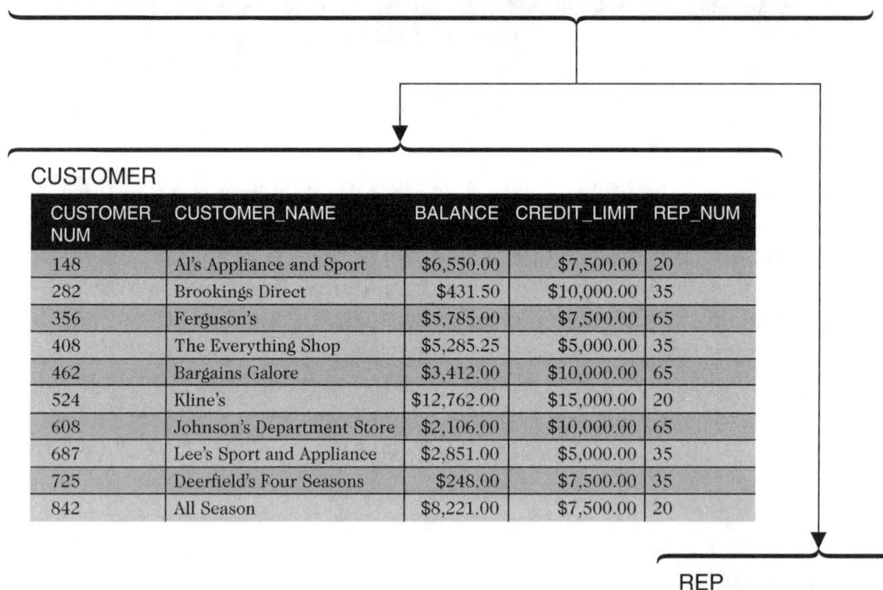

CUSTOMER

CUSTOMER_NUM	CUSTOMER_NAME	BALANCE	CREDIT_LIMIT	REP_NUM
148	Al's Appliance and Sport	$6,550.00	$7,500.00	20
282	Brookings Direct	$431.50	$10,000.00	35
356	Ferguson's	$5,785.00	$7,500.00	65
408	The Everything Shop	$5,285.25	$5,000.00	35
462	Bargains Galore	$3,412.00	$10,000.00	65
524	Kline's	$12,762.00	$15,000.00	20
608	Johnson's Department Store	$2,106.00	$10,000.00	65
687	Lee's Sport and Appliance	$2,851.00	$5,000.00	35
725	Deerfield's Four Seasons	$248.00	$7,500.00	35
842	All Season	$8,221.00	$7,500.00	20

REP

REP_NUM	LAST_NAME	FIRST_NAME
20	Kaiser	Valerie
35	Hull	Richard
65	Perez	Juan

FIGURE 2-12 CUSTOMER table converted to third normal form

Has this new design for the CUSTOMER table corrected all of the previously identified problems? A sales rep's name appears only once, thus avoiding redundancy and simplifying the process of changing a sales rep's name. This design prohibits a sales rep from having different names in the database. To add a new sales rep to the database, you add a row to the REP table; it is not necessary for a new rep to represent a customer. Finally, deleting all customers of a given sales rep will not remove the sales rep's record from the REP table, retaining the sales rep's name in the database. You can reconstruct all the data in the original table from the data in the new collection of tables. All previously mentioned problems have indeed been solved.

Q & A

Question: Convert the following table to third normal form. In this table, STUDENT_NUM determines STUDENT_NAME, NUM_CREDITS, ADVISOR_NUM, and ADVISOR_NAME. ADVISOR_NUM determines ADVISOR_NAME. COURSE_NUM determines DESCRIPTION. The combination of a STUDENT_NUM and a COURSE_NUM determines GRADE.

```
STUDENT (STUDENT_NUM, STUDENT_NAME, NUM_CREDITS,
     ADVISOR_NUM, ADVISOR_NAME, (COURSE_NUM, DESCRIPTION,
     GRADE) )
```

Answer: Complete the following steps:

Step 1. Remove the repeating group to convert the table to first normal form, as follows:

```
STUDENT (STUDENT_NUM, STUDENT_NAME, NUM_CREDITS,
     ADVISOR_NUM, ADVISOR_NAME, COURSE_NUM, DESCRIPTION,
     GRADE)
```

The STUDENT table is now in first normal form because it has no repeating groups. It is not, however, in second normal form because STUDENT_NAME is dependent only on STUDENT_NUM, which is only a portion of the primary key.

Step 2. Convert the STUDENT table to second normal form. First, for each subset of the primary key, start a table with that subset as its key yielding the following:

```
(STUDENT_NUM,
(COURSE_NUM,
(STUDENT_NUM, COURSE_NUM,
```

Next, place the rest of the columns with the smallest collection of columns on which they depend, as follows:

```
(STUDENT_NUM, STUDENT_NAME, NUM_CREDITS, ADVISOR_NUM,
     ADVISOR_NAME)
(COURSE_NUM, DESCRIPTION)
(STUDENT_NUM, COURSE_NUM, GRADE)
```

Finally, assign names to each of the new tables:

```
STUDENT (STUDENT_NUM, STUDENT_NAME, NUM_CREDITS,
     ADVISOR_NUM, ADVISOR_NAME)
COURSE (COURSE_NUM, DESCRIPTION)
STUDENT_COURSE (STUDENT_NUM, COURSE_NUM, GRADE)
```

These tables are all now in second normal form, and the COURSE and STUDENT_COURSE tables are also in third normal form. The STUDENT table is not in third normal form, however, because it contains a determinant (ADVISOR_NUM) that is not a candidate key.

continued

Step 3: Convert the STUDENT table to third normal form by removing the column that depends on the determinant ADVISOR_NUM and placing it in a separate table, as follows:

```
(STUDENT_NUM, STUDENT_NAME, NUM_CREDITS, ADVISOR_NUM)
(ADVISOR_NUM, ADVISOR_NAME)
```

Step 4: Name the tables and put the entire collection together, as follows:

```
STUDENT (STUDENT_NUM, STUDENT_NAME, NUM_CREDITS,
    ADVISOR_NUM)
ADVISOR (ADVISOR_NUM, ADVISOR_NAME)
COURSE (COURSE_NUM, DESCRIPTION)
STUDENT_COURSE (STUDENT_NUM, COURSE_NUM, GRADE)
```

DIAGRAMS FOR DATABASE DESIGN

For many people, an illustration of a database's structure is quite useful. A popular type of illustration used to represent the structure of a database is the **entity-relationship (E-R) diagram**. In an E-R diagram, a rectangle represents an entity (table). One-to-many relationships between entities are drawn as lines between the corresponding rectangles.

Several different styles of E-R diagrams are used to diagram a database design. In the version shown in Figure 2-13, an arrowhead indicates the "many" side of the relationship between tables. In the relationship between the REP and CUSTOMER tables, for example, the arrow points from the REP table to the CUSTOMER table, indicating that one sales rep is related to many customers. The ORDER_LINE table has two one-to-many relationships, as indicated by the line from the ORDERS table to the ORDER_LINE table and the line from the PART table to the ORDER_LINE table.

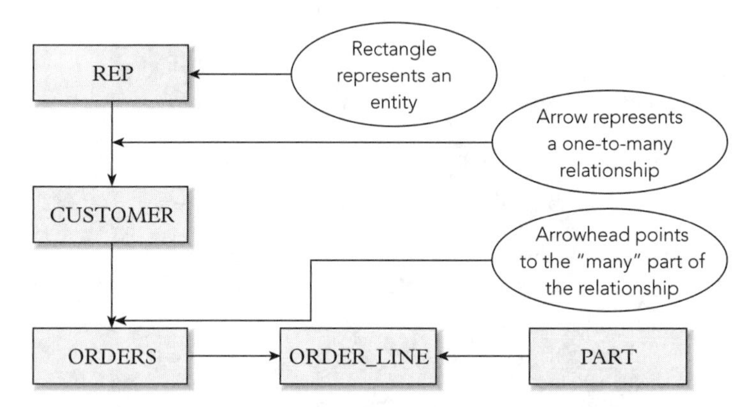

FIGURE 2-13 E-R diagram for the Premiere Products database with rectangles and arrows

NOTE

In this style of E-R diagram, you can put the rectangles in any position to represent the entities and relationships. The important thing is that the arrows connect the appropriate rectangles.

Another style of E-R diagram is to represent the "many" side of a relationship between tables with a crow's foot, as shown in Figure 2-14.

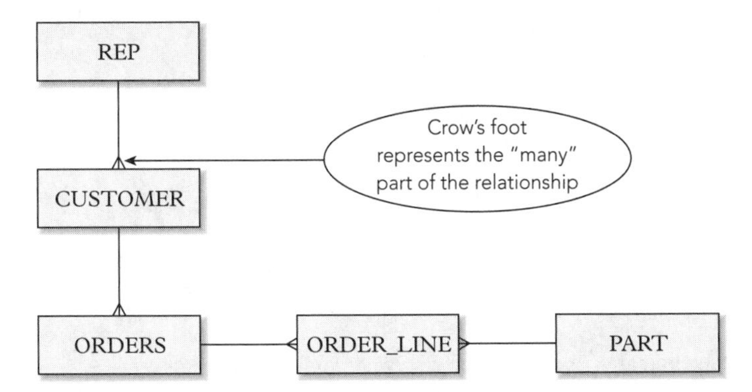

FIGURE 2-14 E-R diagram for the Premiere Products database with a crow's foot

The E-R diagram shown in Figure 2-15 represents the original style of E-R diagrams. In this style, relationships are indicated in diamonds that describe the relationship. The relationship between the REP and CUSTOMER tables, for example, is named REPRESENTS, reflecting the fact that a sales rep represents a customer. The relationship between the CUSTOMER and ORDERS table is named PLACED, reflecting the fact that customers place orders. The relationship between the ORDERS and ORDER_LINE tables is named CONTAINS, reflecting the fact that an order contains order lines. The relationship between the PART and ORDER_LINE tables is named IS_ON, reflecting the fact that a given part is on many orders. In this style of E-R diagram, the number 1 indicates the "one" side of the relationship and the letter "n" represents the "many" side of the relationship.

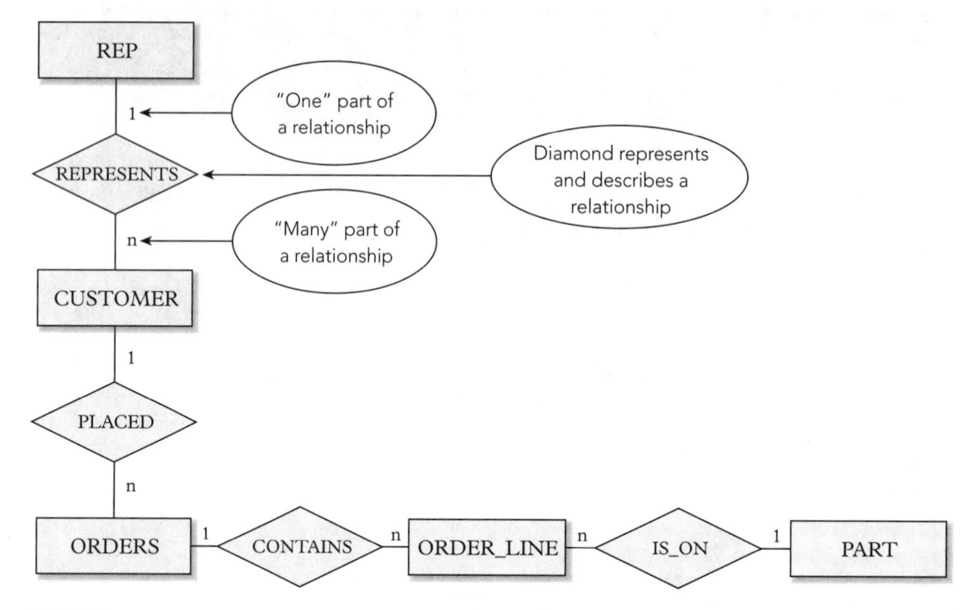

FIGURE 2-15 E-R diagram for the Premiere Products database with named relationships

Chapter Summary

- An entity is a person, place, thing, or event. An attribute is a property of an entity. A relationship is an association between entities.
- A relation is a two-dimensional table in which the entries in the table contain only single values, each column has a distinct name, all values in a column match this name, the order of the rows and columns is immaterial, and each row contains unique values. A relational database is a collection of relations.
- Column B is functionally dependent on another column, A (or possibly a collection of columns), when a value for A determines a single value for B at any one time.
- Column A (or a collection of columns) is the primary key for a relation (table), R, if *all* columns in R are functionally dependent on A and no subcollection of the columns in A (assuming A is a collection of columns and not just a single column) also has property 1.
- To design a database to satisfy a particular set of requirements, first read through the requirements and identify the entities (objects) involved. Give names to the entities and identify the unique identifiers for these entities. Next, identify the attributes for all the entities and the functional dependencies that exist among the attributes, and then use the functional dependencies to identify the tables and columns. Finally, identify any relationships between tables by looking at matching columns.
- A table (relation) is in first normal form (1NF) when it does not contain a repeating group. To convert an unnormalized table to first normal form, remove the repeating group and expand the primary key to include the original primary key along with the key to the repeating group.
- A table (relation) is in second normal form (2NF) when it is in first normal form and no nonkey column (that is, a column that is not part of the primary key) is dependent on only a portion of the primary key. To convert a table in first normal form to a collection of tables in second normal form, take each subset of the set of columns that make up the primary key, and begin a new table with this subset as its primary key. Next, place each of the other columns with the appropriate primary key; that is, place each one with the minimal collection of columns on which it depends. Finally, give each of these new tables a name that is descriptive of the meaning and contents of the table.
- A table is in third normal form (3NF) when it is in second normal form and the only determinants (columns on which at least one other column depends) it contains are candidate keys (columns that could function as the primary key). To convert a table in second normal form to a collection of tables in third normal form, first, for each determinant that is not a candidate key, remove from the table the columns that depend on this determinant (but don't remove the determinant). Next, create a new table containing all the columns from the original table that depend on this determinant. Finally, make the determinant the primary key of this new table.
- An entity-relationship (E-R) diagram is an illustration that represents the design of a database. There are several common styles of illustrating database design that use shapes to represent entities and connectors to illustrate the relationships between those entities.

Key Terms

attribute
Boyce-Codd normal form (BCNF)
candidate key
concatenation
database design
determinant
entity
entity-relationship (E-R) diagram
field
first normal form (1NF)
functionally dependent
functionally determine
nonkey column
normal form
normalization
one-to-many relationship
primary key
qualify
record
redundancy
relation
relational database
relationship
repeating group
second normal form (2NF)
third normal form (3NF)
tuple
unnormalized relation
update anomaly

Review Questions

1. What is an entity?
2. What is an attribute?
3. What is a relationship? What is a one-to-many relationship?
4. What is a repeating group?
5. What is a relation?
6. What is a relational database?
7. Describe the shorthand representation of the structure of a relational database. Illustrate this technique by representing the database for Henry Books as shown in Figures 1-4 through 1-7 in Chapter 1.
8. How do you qualify the name of a field, and when do you need to do this?
9. What does it mean for a column to be functionally dependent on another column?
10. What is a primary key? What is the primary key for each of the tables in the Henry Books database shown in Chapter 1?
11. A database at a college must support the following requirements:
    a. For a department, store its number and name.
    b. For an advisor, store his or her number, last name, first name, and the department number to which the advisor is assigned.
    c. For a course, store its code and description (for example, MTH110, Algebra).
    d. For a student, store his or her number, first name, and last name. For each course the student takes, store the course code, the course description, and the grade earned.

Also, store the number and name of the student's advisor. Assume that an advisor might advise any number of students but that each student has just one advisor.

Design the database for the preceding set of requirements. Use your own experience as a student to determine any functional dependencies. List the tables, columns, and relationships. In addition, represent your design with an E-R diagram.

12. Define first normal form.
13. Define second normal form. What types of problems might you encounter using tables that are not in second normal form?
14. Define third normal form. What types of problems might you encounter using tables that are not in third normal form?
15. Using the functional dependencies you determined in Question 11, convert the following table to an equivalent collection of tables that are in third normal form.

```
STUDENT (STUDENT_NUM, STUDENT_LAST_NAME, STUDENT_FIRST_NAME,
         ADVISOR_NUM, ADVISOR_LAST_NAME, ADVISOR_FIRST_NAME,
         (COURSE_CODE, DESCRIPTION, GRADE) )
```

Exercises

Premiere Products

Answer each of the following questions using the Premiere Products data shown in Figure 2-1. No computer work is required.

1. Indicate the changes (using the shorthand representation) that you would need to make to the original Premiere Products database design (see Figure 2-1) to support the following requirements. A customer is not necessarily represented by a single sales rep, but can be represented by several sales reps. When a customer places an order, the sales rep who gets the commission on the order must be in the collection of sales reps who represent the customer.
2. Indicate the changes (using the shorthand representation) that you would need to make to the original Premiere Products database design to support the following requirements. There is no relationship between customers and sales reps. When a customer places an order, any sales rep can process the order. On the order, you need to identify both the customer placing the order and the sales rep responsible for the order. Draw an E-R diagram for the new design.
3. Indicate the changes (using the shorthand representation) that you would need to make to the original Premiere Products database design in the event that the original Requirement 3 is changed as follows. For a part, store the part's number, description, item class, and price. In addition, for each warehouse in which the part is located, store the number of the warehouse, the description of the warehouse, and the number of units of the part stored in the warehouse. Draw an E-R diagram for the new design.

4. Using your knowledge of Premiere Products, determine the functional dependencies that exist in the following table. After determining the functional dependencies, convert this table to an equivalent collection of tables that are in third normal form.

```
PART (PART_NUM, DESCRIPTION, ON_HAND, CLASS, WAREHOUSE,
      PRICE, (ORDER_NUM, ORDER_DATE, CUSTOMER_NUM,
      CUSTOMER_NAME, NUM_ORDERED, QUOTED_PRICE) )
```

Henry Books

Answer each of the following questions using the Henry Books data shown in Figures 1-4 through 1-7 in Chapter 1. No computer work is required.

1. Ray Henry is considering expanding the activities at his book stores to include movies. He has some ideas for how he wants to do this and he needs you to help with database design activities to address these ideas. In particular, he would like you to design a database for him. He is interested in movies and wants to store information about movies, stars, and directors in a database. He needs to be able to satisfy the following requirements:
   a. For each director, list his or her number, name, the year he or she was born, and the year of death if he or she is deceased.
   b. For each movie, list its number, title, the year the movie was made, and its type.
   c. For each movie, list its number, title, the number and name of its director, the critics' rating, the MPAA rating, the number of awards for which the movie was nominated, and the number of awards the movie won.
   d. For each movie star, list his or her number, name, birthplace, the year he or she was born, and the year of death if he or she is deceased.
   e. For each movie, list its number and title, along with the number and name of all the stars who appear in it.
   f. For each movie star, list his or her number and name, along with the number and name of all the movies in which he or she stars.

   List the tables, columns, and relationships. In addition, represent your design with an E-R diagram.
2. Determine the functional dependencies that exist in the following table, and then convert this table to an equivalent collection of tables that are in third normal form.

```
BOOK (BOOK_CODE, TITLE, TYPE, PRICE (AUTHOR_NUM,
     AUTHOR_LAST, AUTHOR_FIRST) )
```

3. Determine the functional dependencies that exist in the following table, and then convert this table to an equivalent collection of tables that are in third normal form.

```
BOOK (BOOK_CODE, TITLE, TYPE, PRICE, PUB_CODE,
     PUBLISHER_NAME, CITY)
```

Alexamara Marina Group

Answer each of the following questions using the Alexamara Marina Group data shown in Figures 1-8 through 1-12 in Chapter 1. No computer work is required.

1. Design a database that can satisfy the following requirements:
   a. For each marina, list the number, name, address, city, state, and zip code.

b. For each boat owner, list the number, last name, first name, address, city, state, and zip code.

c. For each marina, list all the slips in the marina. For each slip, list the length of the slip, annual rental fee, name and type of the boat occupying the slip, and boat owner's number, last name, and first name.

d. For each possible service category, list the category number and description. In addition, for each service request in a category, list the marina number and slip number for the boat receiving the service, estimated hours for the service, hours already spent on the service, and next date that is scheduled for the particular service.

e. For each service request, list the marina number, slip number, category description, description of the particular service, and a description of the current status of the service.

List the tables, columns, and relationships. In addition, represent your design with an E-R diagram.

2. Determine the functional dependencies that exist in the following table, and then convert this table to an equivalent collection of tables that are in third normal form.

```
MARINA (MARINA_NUM, NAME, (SLIP_NUM, LENGTH, RENTAL_FEE,
        BOAT_NAME) )
```

3. Determine the functional dependencies that exist in the following table, and then convert this table to an equivalent collection of tables that are in third normal form.

```
MARINA_SLIP (SLIP_ID, MARINA_NUM, SLIP_NUM, LENGTH, RENTAL_FEE,
        BOAT_NAME, BOAT_TYPE, OWNER_NUM, LAST_NAME,
        FIRST_NAME)
```

CHAPTER 3

CREATING TABLES

LEARNING OBJECTIVES

Objectives

- Create and run SQL commands
- Create tables
- Identify and use data types to define columns in tables
- Understand and use nulls
- Add rows to tables
- View table data
- Correct errors in a table
- Save SQL commands to a file
- Describe a table's layout using SQL

INTRODUCTION

You already might be an experienced user of a database management system (DBMS). You might find a DBMS at your school's library, at a site on the Internet, or in any other place where you retrieve data using a computer. In this chapter, you will begin your study of **Structured Query Language (SQL)**, which is one of the most popular and widely used languages for retrieving and manipulating database data.

In the mid-1970s, SQL was developed as the data manipulation language for IBM's prototype relational model DBMS, System R, under the name SEQUEL at IBM's San Jose research facilities. In 1980, the language was renamed SQL (but still pronounced "sequel" although the equally popular pronunciation of "S-Q-L" ["ess-cue-ell"] is used in this text) to avoid confusion with an unrelated hardware product named SEQUEL. Most DBMSs use a version of SQL as their data manipulation language.

In this chapter, you will learn the basics of working in SQL. You will learn how to create tables and assign data types to columns. You also will learn about a special type of value, called a null value, and learn how to manage these values in tables. You will learn how to insert data into your tables after you create them. Finally, you will learn how to describe a table's layout using SQL.

CREATING AND RUNNING SQL COMMANDS

You accomplish tasks in SQL by creating and running commands. In order to do so, you need to use a DBMS that supports SQL. This text uses Oracle as the DBMS in which to create and run the commands. The text also indicates differences you will find if you are using Microsoft Access or Microsoft SQL Server 2005. (If you are using MySQL, contact Cengage Learning for the latest edition of *A Guide to MySQL*, by Pratt and Last.)

Although the version of Oracle used in this text is the Oracle Database 10*g* Express Edition, the commands used in this text will work the same in any other version of Oracle. You use the Oracle Database Express Edition by downloading it from the Oracle Web site, installing it, and then starting it using the Microsoft Internet Explorer Web browser.

Starting the Oracle Database Express Edition

After installing the Oracle Database Express Edition, you start it by clicking the Start button, pointing to All Programs, clicking Oracle Database 10*g* Express Edition, and then clicking Go To Database Home Page. Internet Explorer will start and load the home page, which requests your username and password. (Ask your instructor which username and password to use, or use the one you specified when you installed the software. If a different Web browser starts, ask your instructor for help. Other Web browsers might not fully support the examples used in this text.) After entering this information, click the Login button. Figure 3-1 shows the Oracle Database Express Edition home page. You click the icons on the home page to access the various tools. In this text, you will use the SQL tool. The other tools let you administer a database, work with database objects, and run different database utilities. (These features are beyond the scope of this book.)

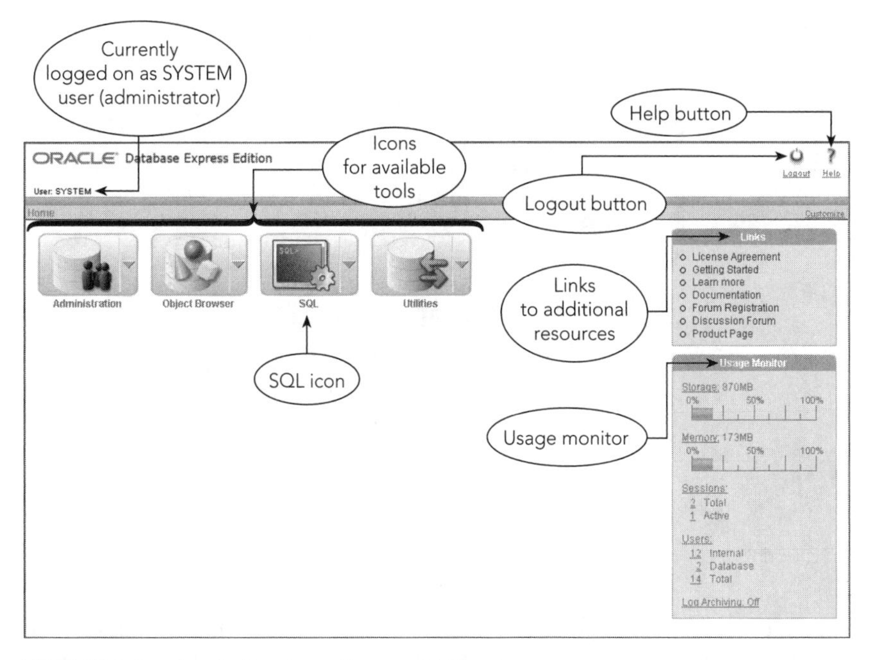

FIGURE 3-1 Oracle Database Express Edition home page

NOTE

The figure shows that the SYSTEM (administrator) user is currently logged on. Your instructor might assign you another username to use as your login, in which case, this name will appear on your screen.

There are two ways to use the tools in the Oracle Database Express Edition. You can click the arrow for the icon to display a menu and then select an option from the menu. Figure 3-2 shows the result of clicking the arrow for the SQL icon and then pointing to the SQL Scripts option on the SQL menu. A submenu of commands for working with SQL scripts appears. To create a script using this approach, for example, you would click the arrow, point to SQL Scripts, and then click Create. (You will learn more about scripts later in this chapter.)

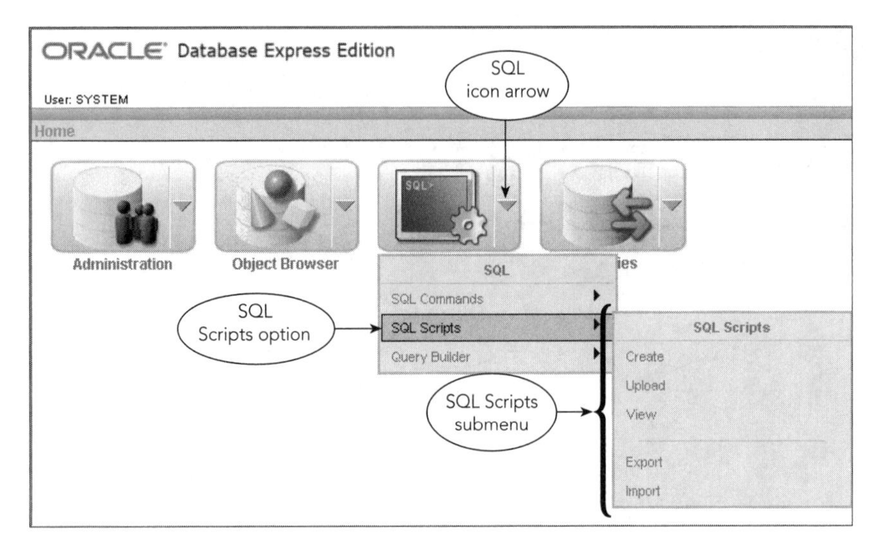

FIGURE 3-2 SQL Scripts submenu

You also can click the icon to display the options as icons instead of as submenus. For example, when you click the SQL icon on the home page, you will see the SQL page shown in Figure 3-3. Clicking an arrow on an icon displays a submenu. The figure shows the results of clicking the arrow for the SQL Scripts icon, which displays the SQL Scripts submenu. To create a script using this approach, click the SQL icon on the home page, click the arrow for the SQL Scripts icon, and then click Create. The approach you choose is a matter of personal preference.

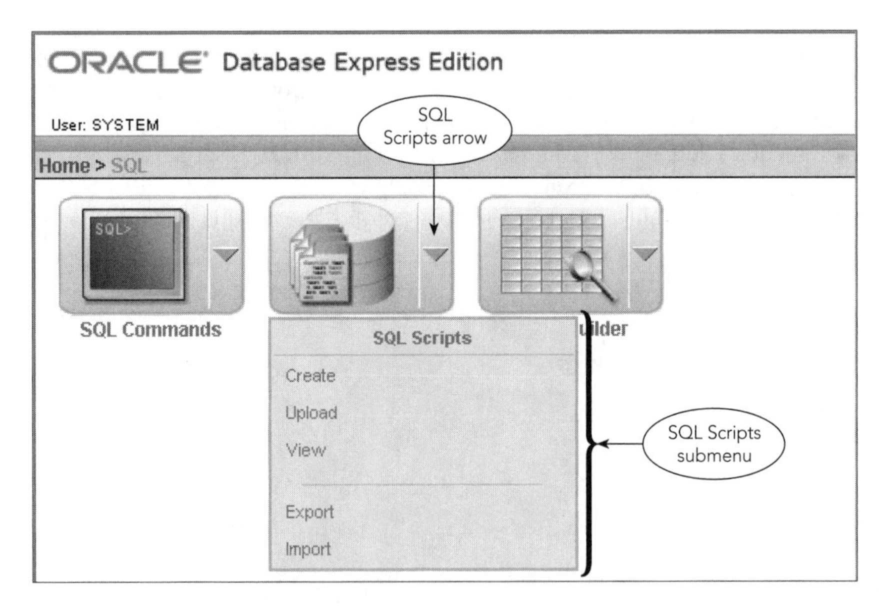

FIGURE 3-3 SQL Scripts submenu on the SQL page

Entering Commands

You enter commands on the SQL Commands page. To access the SQL Commands page, click the arrow for the SQL icon, and then point to SQL Commands as shown in Figure 3-4.

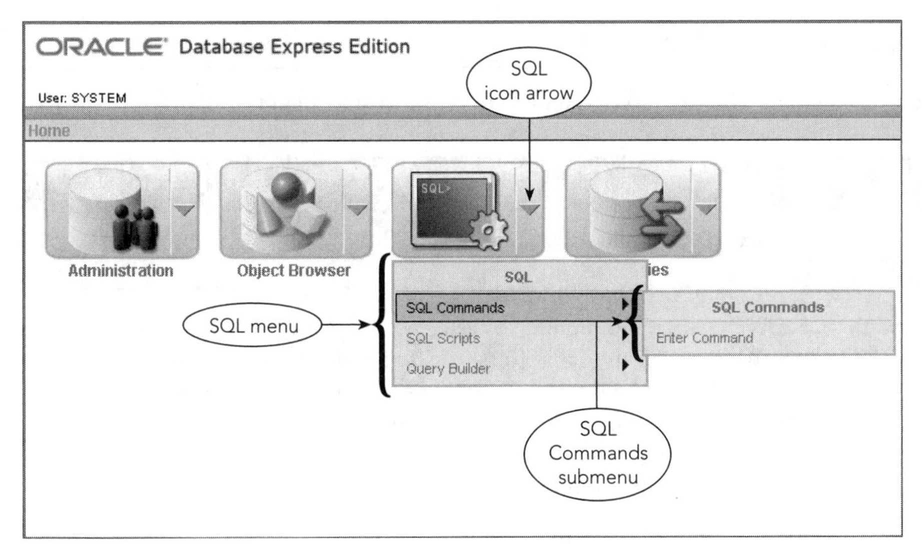

FIGURE 3-4 Starting a new SQL command

After clicking the Enter Command option on the SQL Commands submenu, you will see the SQL Commands page shown in Figure 3-5. You enter the command in the upper portion of this page, called the SQL editor pane, and then click the Run button to execute the command and display its results in the lower portion of the page, called the Results pane.

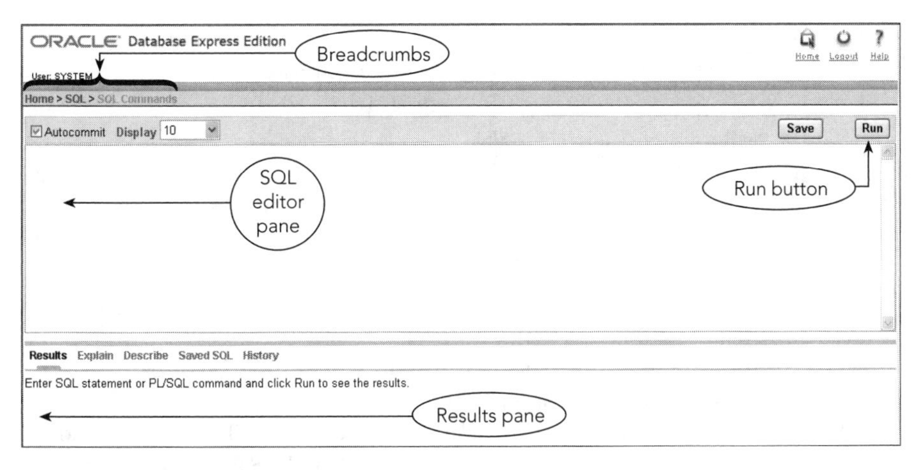

FIGURE 3-5 SQL Commands page

After clicking the Run button, the results will appear in the Results pane. Notice the Home> SQL> SQL Commands reference at the top of the SQL editor pane. This reference is called a **breadcrumb**. You can click the pages in the breadcrumb to move back one or more pages. For example, to return to the home page, click Home in the breadcrumb.

CREATING A TABLE

Before you begin adding data to a table, you must describe the layout of the table to the DBMS.

EXAMPLE 1

Describe the layout of the REP table to the DBMS.

You use the **CREATE TABLE** command to describe the layout of a table. The word TABLE is followed by the name of the table to be created and then by the names and data types of the columns that the table contains. The **data type** indicates the type of data that the column can contain (for example, characters, numbers, or dates) as well as the maximum number of characters or digits that the column can store.

The restrictions placed on table and column names are as follows:

1. The names cannot exceed 30 characters.
2. The names must start with a letter.
3. The names can contain letters, numbers, and underscores (_).
4. The names cannot contain spaces.

The SQL command that creates the REP table is shown in Figure 3-6.

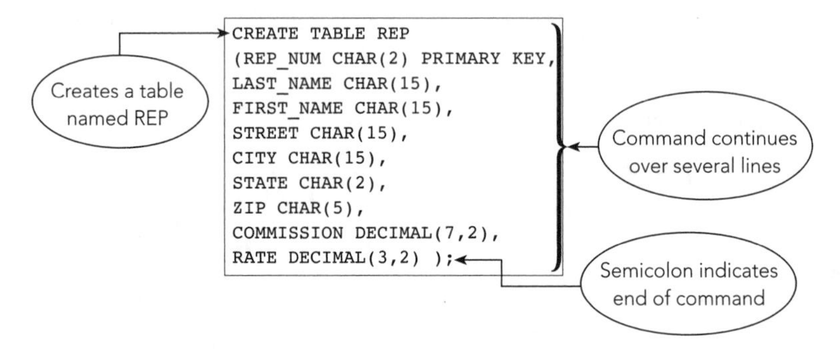

FIGURE 3-6 CREATE TABLE command for the REP table

This CREATE TABLE command, which uses the data definition features of SQL, describes a table named REP. The table contains nine columns: REP_NUM, LAST_NAME, FIRST_NAME, STREET, CITY, STATE, ZIP, COMMISSION, and RATE. The REP_NUM column can store two characters and is the table's primary key. The LAST_NAME column can store 15 characters, and the STATE column can store two characters. The COMMISSION column can store only numbers, and those numbers are limited to seven digits, including two decimal places. Similarly, the RATE column can store three numbers, including two decimal places. You can think of the SQL command shown in Figure 3-6 as creating an empty table with column headings for each column name.

In SQL, commands are free format; that is, no rule says that a particular word must begin in a particular position on the line. For example, you could have written the CREATE TABLE command shown in Figure 3-6 as follows:


```
CREATE TABLE REP (REP_NUM CHAR(2) PRIMARY KEY, LAST_NAME
CHAR(15), FIRST_NAME CHAR(15), STREET CHAR(15), CITY
CHAR(15), STATE CHAR(2), ZIP CHAR(5), COMMISSION
DECIMAL(7,2), RATE DECIMAL(3,2) );
```

The manner in which the CREATE TABLE command shown in Figure 3-6 was written makes the command more readable. This text will strive for such readability when writing SQL commands.

NOTE

SQL is not case sensitive; you can type commands using uppercase or lowercase letters. There is one exception to this rule, however. When you are inserting character values into a table, you must use the correct case.

To create the REP table in Oracle, click in the SQL editor pane, type the CREATE TABLE command shown in Figure 3-7, and then click the Run button on the right side of the SQL editor pane to execute the command and create the table. Figure 3-7 also shows the message that appears in the Results pane after running the command, which indicates that the table was created.

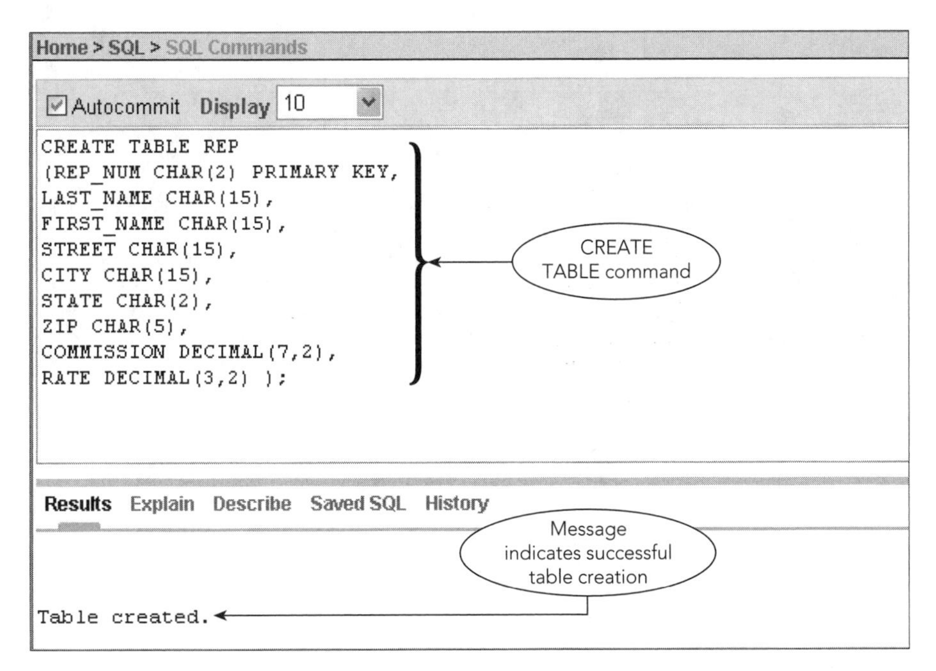

FIGURE 3-7 Running the CREATE TABLE command for the REP table

ACCESS USER NOTE

Microsoft Office Access is a DBMS that lets you work in a graphical user interface, but you can also use it to run SQL commands. To run SQL commands in Access, you must first create a new query. In Access 2007, open the database, click the Create tab on the Ribbon, click the Query Design button in the Other group, close the Show Table dialog box, and then click the SQL View button in the Results group. (In Access 2003, open the database, and then create a new query in Design view. Close the Show Table dialog box, and then click the SQL View button on the Query Design toolbar.) In the Query window, type the SQL command, and then click the Run button on the Query Design toolbar (Access 2003) or the Run button in the Results group on the Design tab (Access 2007) to execute the command. Figure 3-8 shows the command to create the REP table using Access 2007. When you click the Run button, Access will create the table shown in the CREATE TABLE command. (Unlike Oracle, Access doesn't display a message indicating the result was successful.)

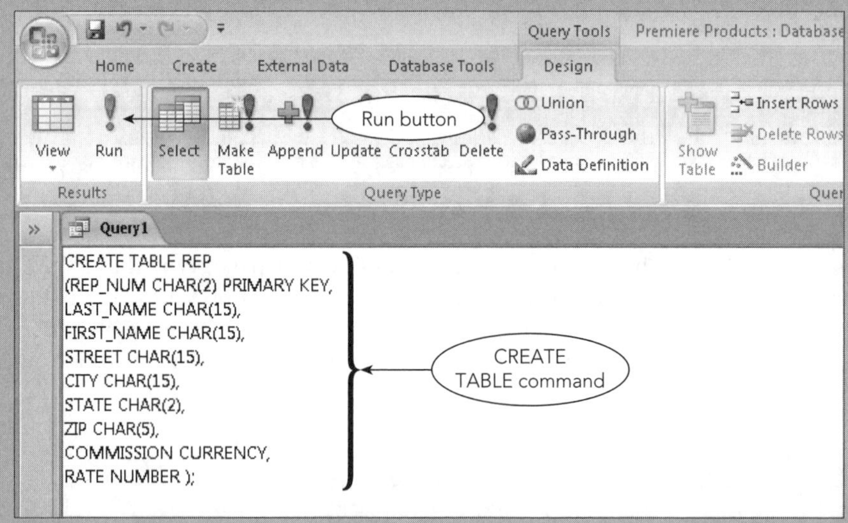

FIGURE 3-8 Using Access SQL view to create a table

Unlike Oracle, Access does not support the DECIMAL data type. To create numbers with decimals, you must use either the CURRENCY or NUMBER data type. Use the CURRENCY data type for fields that will contain currency values; use the NUMBER data type for all other numeric fields.

In Access, it is common to create a table using Table Design view and then to add records to the table using Datasheet view. You still can run SQL commands when you create tables using Design view and enter data into them using Datasheet view.

SQL SERVER USER NOTE

Microsoft SQL Server 2005 is a DBMS designed for use in client-server applications. You can run Microsoft SQL Server 2005 from your own computer through a set of client database tools called SQL Server Management Studio. Management Studio includes a Query Editor window that you can use to run SQL commands. If you are using Management Studio and connecting to a database on your local computer, accept the default values for Server Type, Server Name, and Authentication, and then click the Connect button in the Connect to Server dialog box. When Management Studio appears, double-click Databases, open the database on which you want to run SQL commands, and click the New Query button on the toolbar. Type the SQL command in the Query Editor window that opens, and then click the Execute button on the toolbar to execute the command. The command shown in Figure 3-9 creates the REP table and displays a message in the Messages pane to indicate that the command completed successfully.

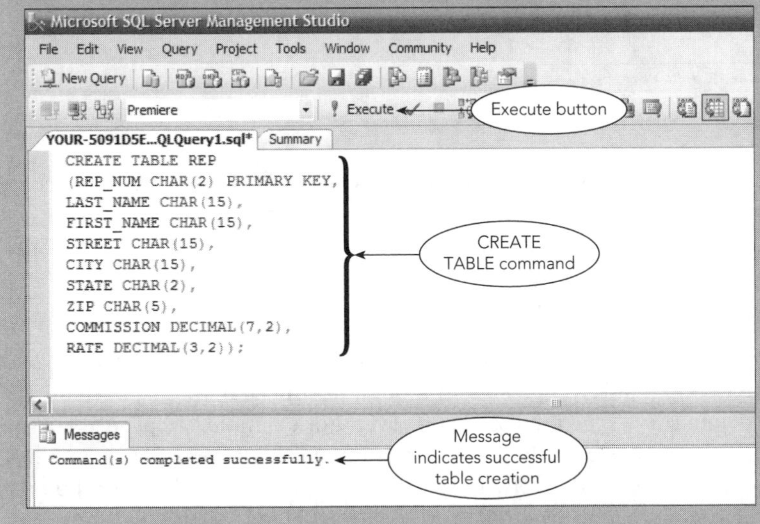

FIGURE 3-9 Using Microsoft SQL Server 2005 to create a table

Correcting Errors in SQL Commands

Suppose that you executed the REP table using the CREATE TABLE command shown in Figure 3-10, which contains several mistakes. Instead of displaying a message that the table was created successfully, Oracle displays an error message about a problem that it encountered. In reviewing the command, you see that CHAR is misspelled on line 4, line 5 is missing a comma, the CITY column was omitted, and line 7 should be deleted. If you run an SQL command and Oracle displays an error, you can use the mouse and the arrow keys on the keyboard to position the insertion point in the correct position so you can correct these errors using the same techniques that you might use in a word processor. For example, you can use the pointer to select the word CHR on line 4 and type CHAR. Then you can use the pointer to move the insertion point to the end of line 5 so you can type the missing comma, and then press Enter to insert the missing information to create the CITY column. You can use the pointer to select the contents of line 7 and then press Delete to remove it. After making these changes, you can click the Run button to execute the command again. If the command contains additional errors, you'll see an error message again. If the command is correct, you'll see the message that the table was created.

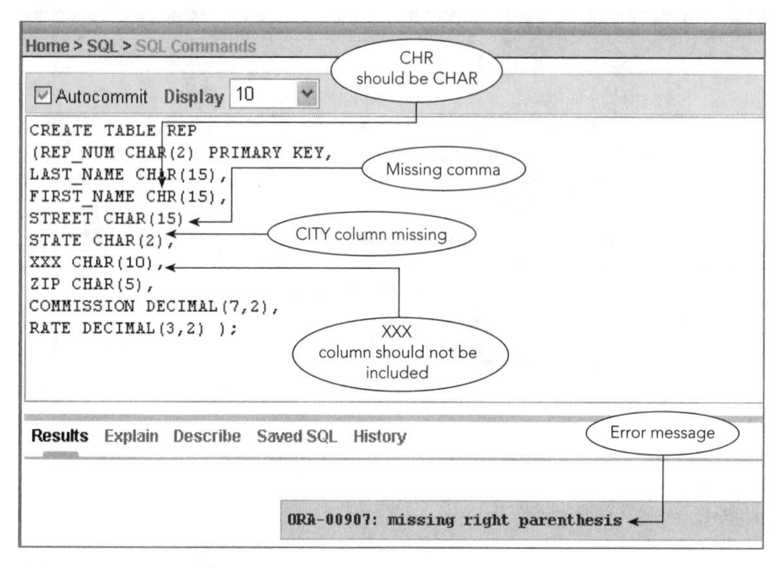

FIGURE 3-10 CREATE TABLE command with errors

Dropping a Table

After creating a table, you might notice that you added a column that you do not need or that you assigned the wrong data type or size to a column. Another way of correcting errors in a table is to delete (drop) the table and start over. For example, suppose you wrote a CREATE TABLE command that contained a column named LST instead of LAST or defined a column as CHAR(5) instead of CHAR(15). Suppose you do not discover the error and you execute the command, creating a table with these problems. In this case, you can delete the entire table using the **DROP TABLE** command and then re-create the table using the correct CREATE TABLE command.

To drop a table, execute the DROP TABLE command, followed by the name of the table you want to delete and a semicolon. To delete the REP table, for example, you would enter the following command and then click the Run button:

```
DROP TABLE REP;
```

Dropping a table also deletes any data that you entered into the table. It is a good idea to check your CREATE TABLE commands carefully before executing them and to correct any problems before adding data. Later in this text, you will learn how to change a table's structure without having to delete the entire table.

Q & A

Question: How can I correct a mistake that I made when I created a table?
Answer: Later in the text, you will see how to alter a table to make any necessary corrections. For now, the easiest way is to drop the table using the DROP TABLE command and then to execute the correct CREATE TABLE command.

USING DATA TYPES

For each column in a table, you must specify the **data type** to use to store the type of data that the column will contain. Figure 3-11 describes some common data types used in databases.

Data type	Description
CHAR(*n*)	Stores a character string *n* characters long. You use the CHAR data type for columns that contain letters and special characters and for columns containing numbers that will not be used in any calculations. Because neither sales rep numbers nor customer numbers will be used in any calculations, for example, the REP_NUM and CUSTOMER_NUM columns are both assigned the CHAR data type.
VARCHAR(*n*)	An alternative to CHAR that stores a character string up to *n* characters long. Unlike CHAR, only the actual character string is stored. If a character string 20 characters long is stored in a CHAR(30) column, for example, it will occupy 30 characters (20 characters plus 10 blank spaces). If it is stored in a VARCHAR(30) column, it will only occupy 20 spaces. In general, tables that use VARCHAR instead of CHAR occupy less space, but the DBMS does not process them as rapidly during queries and updates. However, both are legitimate choices. This text uses CHAR, but VARCHAR would work equally well.
DATE	Stores date data. The specific format in which dates are stored varies from one SQL implementation to another. In Oracle, dates are enclosed in single quotation marks and have the format DD-MON-YYYY (for example, '15-OCT-2010' is October 15, 2010). In Access, dates are enclosed in number signs and are entered using the format MM/DD/YYYY (for example, #10/15/2010# is October 15, 2010). In SQL Server, use the DATETIME data type to store dates.
DECIMAL(*p*,*q*)	Stores a decimal number *p* digits long with *q* of these digits being decimal places to the right of the decimal point. For example, the data type DECIMAL(5,2) represents a number with three places to the left and two places to the right of the decimal (for example, 100.00). You can use the contents of DECIMAL columns in calculations. You also can use the NUMBER (*p*,*q*) data type in both Oracle and SQL Server to store a decimal number. Access does not support the DECIMAL data type; use the CURRENCY or NUMBER data type instead.
INT	Stores integers, which are numbers without a decimal part. The valid range is -2147483648 to 2147483647. You can use the contents of INT columns in calculations. If you follow the word INT with AUTO_INCREMENT, you create a column for which SQL will automatically generate a new sequence number each time you add a new row. This would be the appropriate choice, for example, when you want the DBMS to generate a value for a primary key.
SMALLINT	Stores integers, but uses less space than the INT data type. The valid range is -32768 to 32767. SMALLINT is a better choice than INT when you are certain that the column will store numbers within the indicated range. You can use the contents of SMALLINT columns in calculations.

FIGURE 3-11 Commonly used data types

USING NULLS

Occasionally, when you enter a new row into a table or modify an existing row, the values for one or more columns are unknown or unavailable. For example, you can add a customer's name and address to a table even though the customer does not have an assigned sales rep or an established credit limit. In other cases, some values might never be known—perhaps there is a customer that does not have a sales rep. In SQL, you handle this situation by using a special value to represent cases in which an actual value is unknown, unavailable, or not applicable. This special value is called a **null data value**, or simply a **null**. When creating a table, you can specify whether to allow nulls in the individual columns.

Q & A

Question: Should a user be allowed to enter null values for the primary key?
Answer: No. The primary key is supposed to uniquely identify a given row, and this would be impossible if nulls were allowed. For example, if you stored two customer records without values in the primary key column, you would have no way to tell them apart.

In SQL, you use the **NOT NULL** clause in a CREATE TABLE command to indicate columns that *cannot* contain null values. The default is to allow nulls; columns for which you do not specify NOT NULL can accept null values.

For example, suppose that the LAST_NAME and FIRST_NAME columns in the REP table cannot accept null values, but all other columns in the REP table can. The following CREATE TABLE command accomplishes this goal:

```
CREATE TABLE REP
(REP_NUM CHAR(2) PRIMARY KEY,
LAST_NAME CHAR(15) NOT NULL,
FIRST_NAME CHAR(15) NOT NULL,
STREET CHAR(15),
CITY CHAR(15),
STATE CHAR(2),
ZIP CHAR(5),
COMMISSION DECIMAL(7,2),
RATE DECIMAL(3,2) );
```

If you created the REP table with this CREATE TABLE command, the DBMS would reject any attempt to store a null value in either the LAST_NAME or FIRST_NAME column. The system would accept an attempt to store a null value in the STREET column, however, because the STREET column can accept null values. Because the primary key column cannot accept null values, you do not need to specify the REP_NUM column as NOT NULL.

ADDING ROWS TO A TABLE

After you have created a table in a database, you can load data into the table by using the INSERT command.

The INSERT Command

The **INSERT** command adds rows to a table. You type INSERT INTO followed by the name of the table into which you are adding data. Then you type the word VALUES followed by the specific values to be inserted in parentheses. When adding rows to character columns, make sure you enclose the values in single quotation marks (for example, 'Kaiser'). You also must enter the values in the appropriate case, because character data is stored exactly as you enter it.

NOTE

You must enclose values in single quotation marks for any column whose type is character (CHAR), even when the data contains numbers. Because the ZIP column in the REP table has a CHAR data type, for example, you must enclose zip codes in single quotation marks, even though they are numbers.

If you need to enter an apostrophe (single quotation mark) into a column, you type two single quotation marks. For example, to enter the name O'Toole in the LAST_NAME column, you would type 'O''Toole' as the value in the INSERT command.

EXAMPLE 2

Add sales rep 20 to the REP table.

The command for this example is shown in Figure 3-12. Note that the character strings ('20','Kaiser','Valerie', and so on) are enclosed in single quotation marks. When you execute the command, the record is added to the REP table.

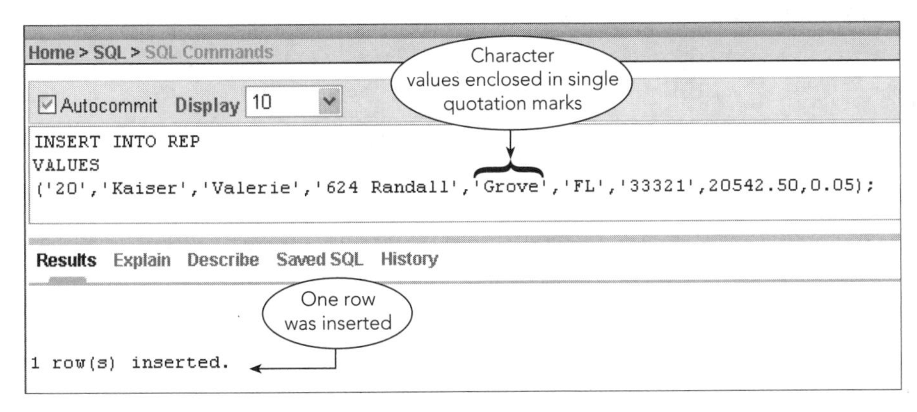

FIGURE 3-12 INSERT command for the first record in the REP table

NOTE

Make sure that you type the values in the same case as those shown in the figures to avoid problems later when retrieving data from the database.

EXAMPLE 3

Add sales reps 35 and 65 to the REP table.

You could enter and execute new INSERT commands to add the new rows to the table. However, an easier and faster way to add these new rows to the table is to use the mouse and the keyboard to modify the previous INSERT command and execute it to add the record for the second sales rep, as shown in Figure 3-13.

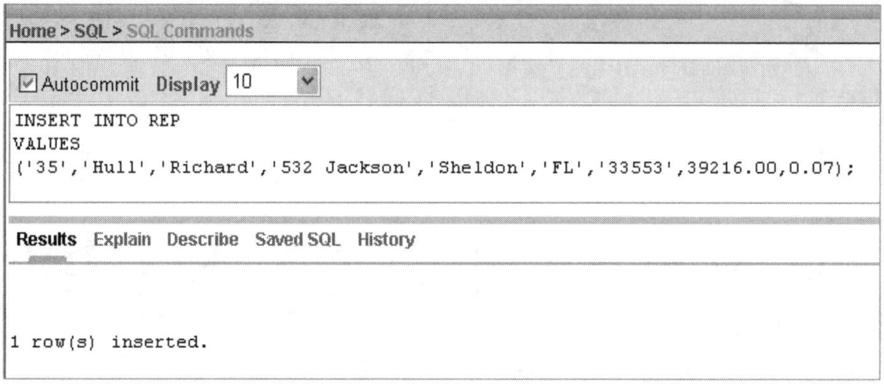

Home > SQL > SQL Commands

Autocommit Display 10

```
INSERT INTO REP
VALUES
('35','Hull','Richard','532 Jackson','Sheldon','FL','33553',39216.00,0.07);
```

Results Explain Describe Saved SQL History

```
1 row(s) inserted.
```

FIGURE 3-13 INSERT command to add the second row to the REP table

You can modify and execute the INSERT command again for the third sales rep, as shown in Figure 3-14, to add the third row to the table.

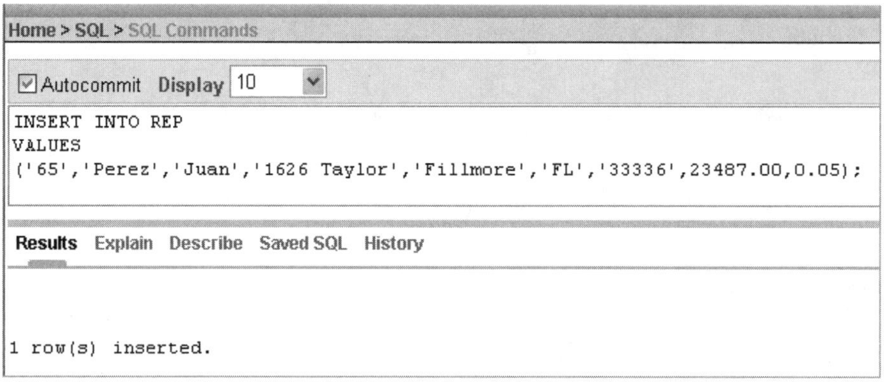

Home > SQL > SQL Commands

Autocommit Display 10

```
INSERT INTO REP
VALUES
('65','Perez','Juan','1626 Taylor','Fillmore','FL','33336',23487.00,0.05);
```

Results Explain Describe Saved SQL History

```
1 row(s) inserted.
```

FIGURE 3-14 INSERT command to add the third row to the REP table

Inserting a Row that Contains Nulls

To enter a null value into a table, you use a special form of the INSERT command in which you identify the names of the columns that will accept non-null values, and then list only these non-null values after the VALUES command, as shown in Example 4.

EXAMPLE 4

Add sales rep 85 to the REP table. Her name is Tina Webb. All columns except REP_NUM, LAST_NAME, and FIRST_NAME are null.

In this case, you do not enter a value of *null*; you enter only the non-null values. To do so, you must indicate precisely which values you are entering by listing the corresponding columns as shown in Figure 3-15. The command shown in Figure 3-15 indicates that you are entering data in only the REP_NUM, LAST_NAME, and FIRST_NAME columns and that you are *not* entering values in any other columns; the other columns will contain null values.

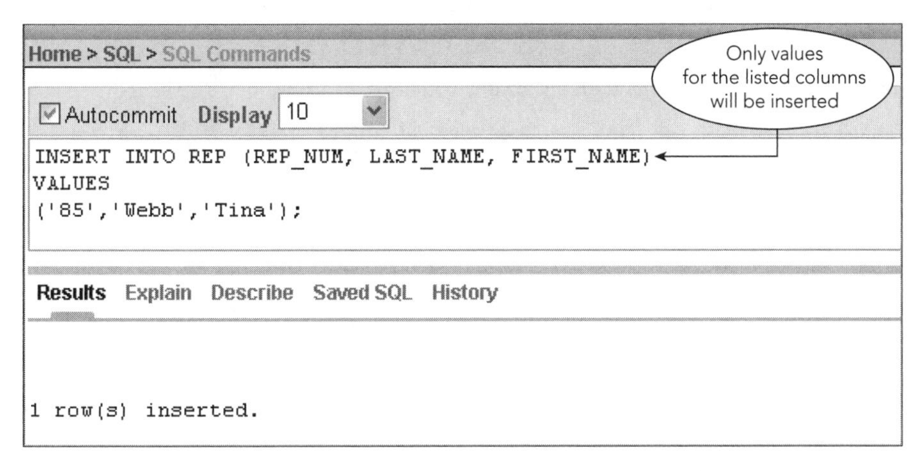

FIGURE 3-15 Inserting a row in the REP table that contains null values

VIEWING TABLE DATA

To view the data in a table, you use the **SELECT** command, which is described in more detail in Chapters 4 and 5.

EXAMPLE 5

Display all the rows and columns in the REP table.

You can use a simple version of the SELECT command to display all the rows and columns in a table by typing the word SELECT, followed by an asterisk, followed by the word FROM and the name of the table containing the data you want to view. Just as with other SQL commands, the command ends with a semicolon. In Oracle, you type the command shown in Figure 3-16, and then click the Run button to display the results.

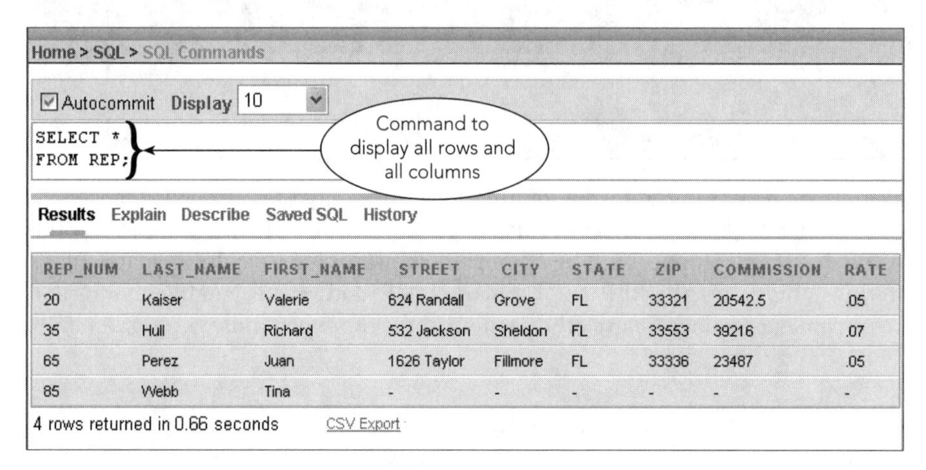

REP_NUM	LAST_NAME	FIRST_NAME	STREET	CITY	STATE	ZIP	COMMISSION	RATE
20	Kaiser	Valerie	624 Randall	Grove	FL	33321	20542.5	.05
35	Hull	Richard	532 Jackson	Sheldon	FL	33553	39216	.07
65	Perez	Juan	1626 Taylor	Fillmore	FL	33336	23487	.05
85	Webb	Tina	-	-	-	-	-	-

FIGURE 3-16 Using a SELECT command to view table data

ACCESS USER NOTE

In Access, type the query shown in Figure 3-17 in SQL view to display all the rows and columns in a table.

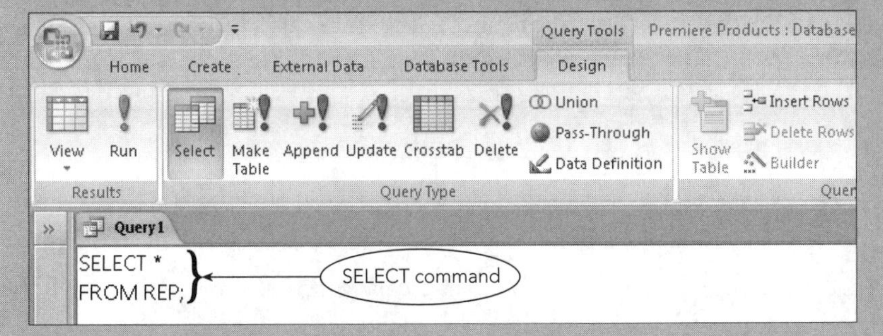

FIGURE 3-17 Using a SELECT command to view table data in Access

To run the query, click the Run button. Access will display the query results in Datasheet view, as shown in Figure 3-18. If the data does not fit on the screen, you can adjust the columns to best fit the data they contain by double-clicking the right edge of each column heading. You can use the scroll bars when necessary to view data that has scrolled off the screen.

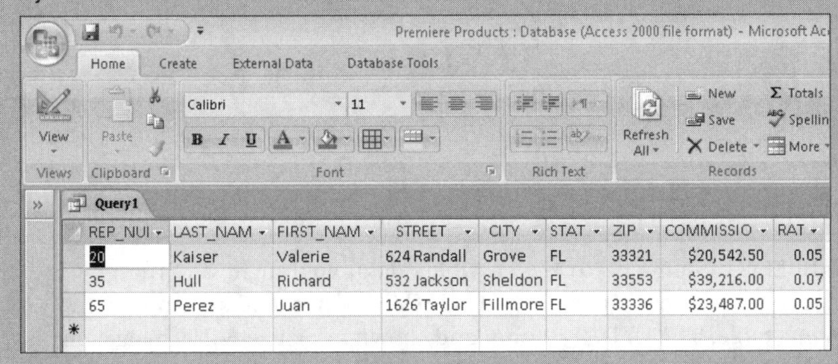

REP_NUI	LAST_NAM	FIRST_NAM	STREET	CITY	STAT	ZIP	COMMISSIO	RAT
20	Kaiser	Valerie	624 Randall	Grove	FL	33321	$20,542.50	0.05
35	Hull	Richard	532 Jackson	Sheldon	FL	33553	$39,216.00	0.07
65	Perez	Juan	1626 Taylor	Fillmore	FL	33336	$23,487.00	0.05

FIGURE 3-18 Query results in Access Datasheet view

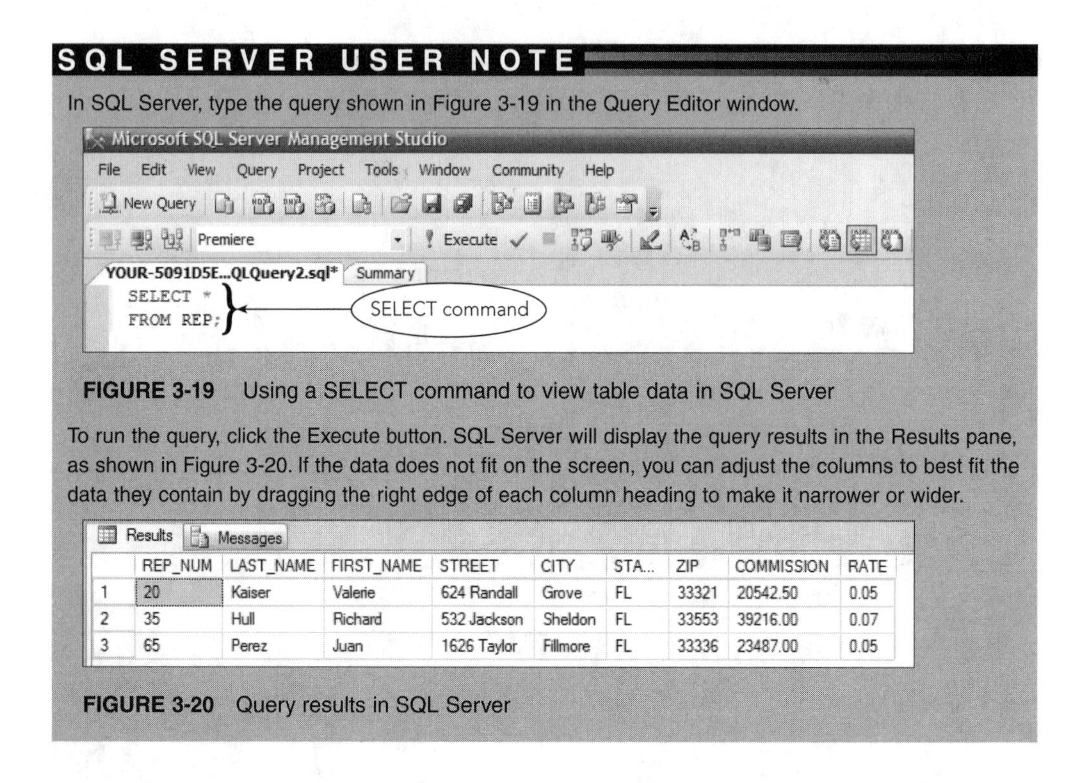

SQL SERVER USER NOTE

In SQL Server, type the query shown in Figure 3-19 in the Query Editor window.

FIGURE 3-19 Using a SELECT command to view table data in SQL Server

To run the query, click the Execute button. SQL Server will display the query results in the Results pane, as shown in Figure 3-20. If the data does not fit on the screen, you can adjust the columns to best fit the data they contain by dragging the right edge of each column heading to make it narrower or wider.

	REP_NUM	LAST_NAME	FIRST_NAME	STREET	CITY	STA...	ZIP	COMMISSION	RATE
1	20	Kaiser	Valerie	624 Randall	Grove	FL	33321	20542.50	0.05
2	35	Hull	Richard	532 Jackson	Sheldon	FL	33553	39216.00	0.07
3	65	Perez	Juan	1626 Taylor	Fillmore	FL	33336	23487.00	0.05

FIGURE 3-20 Query results in SQL Server

CORRECTING ERRORS IN A TABLE

After executing a SELECT command to view a table's data, you might find that you need to change the value in a column. You can use the **UPDATE** command shown in Figure 3-21 to change a value in a table. The UPDATE command shown in Figure 3-21 changes the last name in the row on which the sales rep number is 85 to Perry.

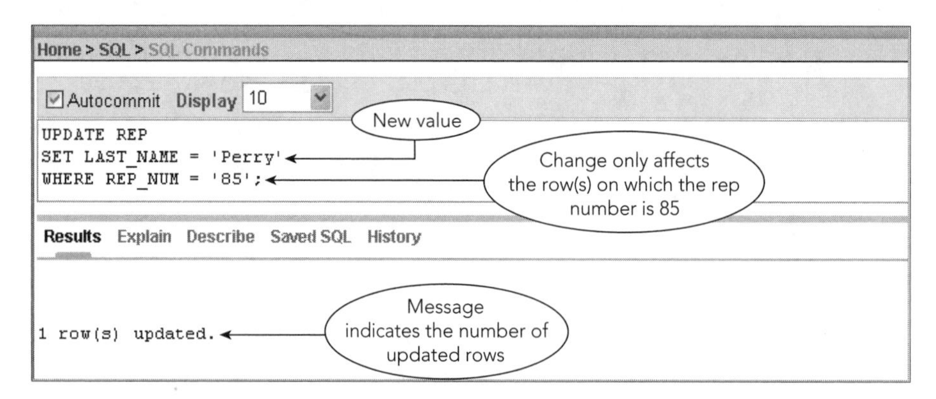

FIGURE 3-21 Using an UPDATE command to change a value

The SELECT command shown in Figure 3-22 displays the results of the UPDATE command shown in Figure 3-21, in which the last name for rep number 85 is Perry.

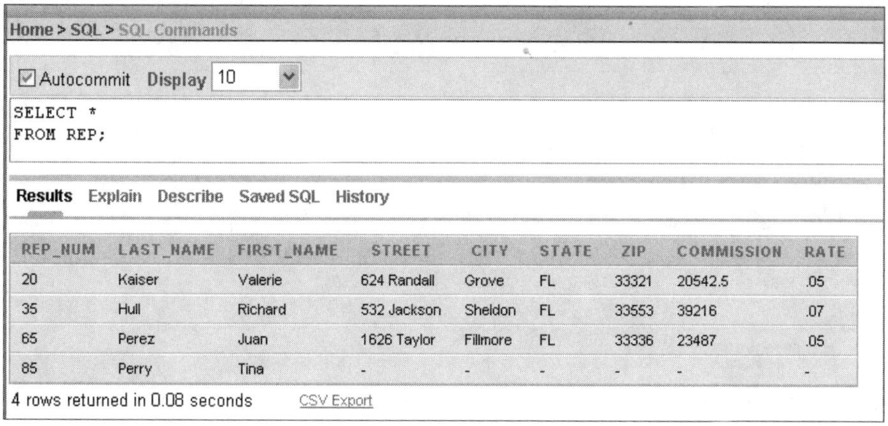

Home > SQL > SQL Commands

Autocommit Display 10

```
SELECT *
FROM REP;
```

Results Explain Describe Saved SQL History

REP_NUM	LAST_NAME	FIRST_NAME	STREET	CITY	STATE	ZIP	COMMISSION	RATE
20	Kaiser	Valerie	624 Randall	Grove	FL	33321	20542.5	.05
35	Hull	Richard	532 Jackson	Sheldon	FL	33553	39216	.07
65	Perez	Juan	1626 Taylor	Fillmore	FL	33336	23487	.05
85	Perry	Tina	-	-	-	-	-	-

4 rows returned in 0.08 seconds CSV Export

FIGURE 3-22 Last name changed for sales rep number 85

When you need to delete a row from a table, you can use the **DELETE** command. The DELETE command shown in Figure 3-23 deletes any row on which the sales rep number is 85.

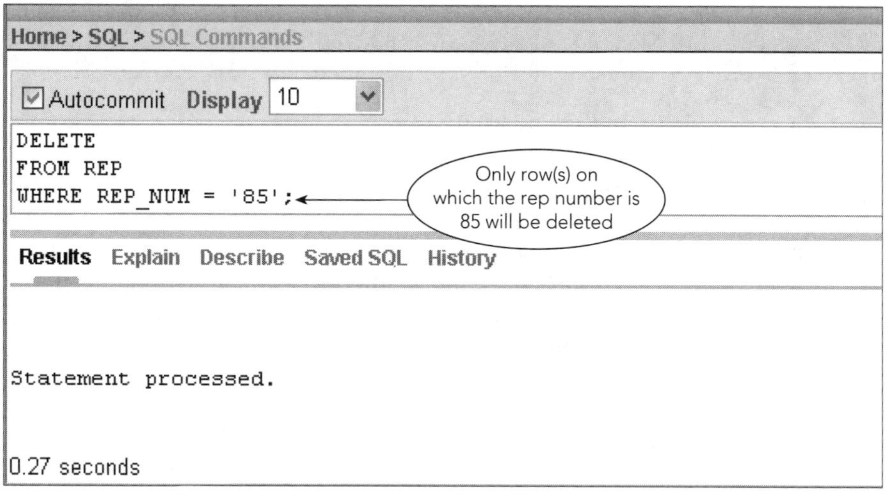

Home > SQL > SQL Commands

Autocommit Display 10

```
DELETE
FROM REP
WHERE REP_NUM = '85';
```

Only row(s) on which the rep number is 85 will be deleted

Results Explain Describe Saved SQL History

```
Statement processed.
```

0.27 seconds

FIGURE 3-23 Using a DELETE command to delete a row

The SELECT command shown in Figure 3-24 displays the updated data.

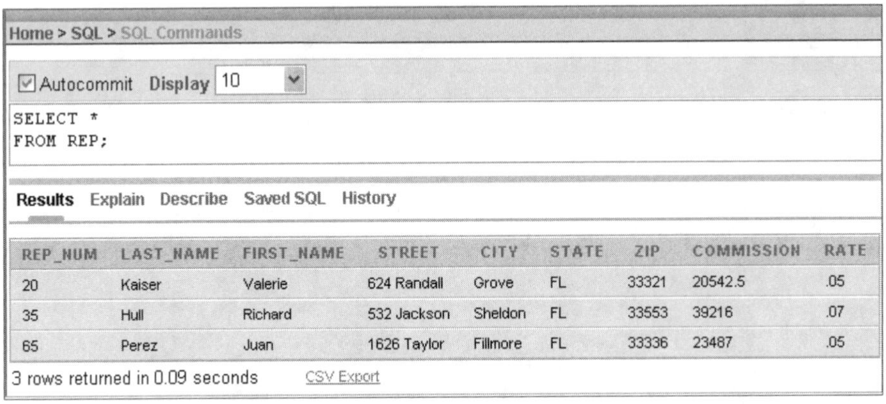

Home > SQL > SQL Commands

Autocommit Display 10

```
SELECT *
FROM REP;
```

Results Explain Describe Saved SQL History

REP_NUM	LAST_NAME	FIRST_NAME	STREET	CITY	STATE	ZIP	COMMISSION	RATE
20	Kaiser	Valerie	624 Randall	Grove	FL	33321	20542.5	.05
35	Hull	Richard	532 Jackson	Sheldon	FL	33553	39216	.07
65	Perez	Juan	1626 Taylor	Fillmore	FL	33336	23487	.05

3 rows returned in 0.09 seconds CSV Export

FIGURE 3-24 Sales rep number 85 deleted from REP table

Q & A

Question: How do I correct errors in my data?
Answer: The method you use to correct an error depends on the type of error you need to correct. If you added a row that should not be in the table, use a DELETE command to remove it. If you forgot to add a row, you can use an INSERT command to add it. If you added a row that contains incorrect data, you can use an UPDATE command to make the necessary corrections. Alternatively, you could use a DELETE command to remove the row containing the error and then use an INSERT command to insert the correct row.

SAVING SQL COMMANDS

Oracle lets you save a command so you can use it again without retyping it. In many DBMSs, you save commands in a **script file**, or simply a **script**, which is a text file with the .sql filename extension. When you use Oracle to create a script, Oracle stores the script in a special location called the **script repository**. If you want to save a script on the local file system, such as on a hard drive or USB drive, you can do so by downloading the script. When you need to use a script that is stored on the local file system, but is not currently stored in Oracle, you can upload the script so you can use it in Oracle. The following steps describe how to create and use scripts in the Oracle Database Express Edition. If you are using a different version of Oracle or another DBMS, use Help or consult the system documentation to determine how to accomplish the same tasks.

To create a script:

1. Load the Oracle Database Express Edition home page and log in.
2. Click the SQL icon arrow, point to SQL Scripts, and then click Create. The Script Editor page opens.
3. In the Script Name text box, type a name for the script.

4. Click in the text box on the page to activate it, and then type the command or commands to save in the script. When necessary, click the Run button to execute the commands saved in the script.
5. When you are finished, click the Save button. You return to the SQL Scripts page and the script you created appears as an icon on the page.

To view, edit, or run an existing script:

1. Load the Oracle Database Express Edition home page and log in.
2. Click the SQL icon arrow, point to SQL Scripts, and then click View.
3. Click the icon for the desired script. The script appears on the Script Editor page. You can use this page to view the content of the script or to make changes to it by editing the commands it contains. If you edit a script, click the Save button to save your changes.
4. To run a script, click the Run button. The Run Script page loads and asks you to confirm running the script. Click the Run button on the Run Script page. The Manage Script Results page opens and displays the script name and an icon in the View Results column. To see the results of the command stored in the script, click the icon in the View Results column.

When you are finished using a script or no longer need to store it, you can delete it. To delete a script:

1. Follow the previous instructions to view the script.
2. Click the Delete button on the Script Editor page.
3. Click the OK button to confirm the deletion.

To download a script from the script repository so you can save it as a file:

1. Follow the previous instructions to view the script.
2. Click the Download button. The File Download dialog box opens.
3. In the dialog box, click the Save button, and then save the file to the desired location.
4. Click the Close button to close the Download complete dialog box.

To upload a script to the script repository:

1. Load the Oracle Database Express Edition home page and log in.
2. Click the SQL icon arrow, point to SQL Scripts, and then click Upload. The Upload Script page opens.
3. Click the Browse button. The Choose file dialog box opens. Navigate to and select the script file to upload. Click the Open button. (If you want to upload the script with a different filename, type the new name in the Script Name text box.)
4. On the Upload Script page, click the Upload button. An icon for the script appears on the SQL Scripts page.

ACCESS USER NOTE

Access does not use script files, but you can save an SQL command as a query object in the database. Open the database, create the query in SQL view, click the Save button on the Quick Access toolbar (or on the menu bar), and then save the query with the desired object name. To run the query without first viewing the SQL command, right-click the query in the Navigation Pane (or in the Database window), and then click Open on the shortcut menu. The query results will appear in Datasheet view. If you want to view the SQL command before running the query, right-click the query in the Navigation Pane, and then click Design View on the shortcut menu. To run the SQL command, click the Run button.

SQL SERVER USER NOTE

SQL Server can store scripts in any folder on your local system. All scripts created in SQL Server are text files with the .sql extension. To create a script file in SQL Server:

1. Load SQL Server Management Studio, and then click the Connect button in the Connect to Server dialog box.
2. Open the appropriate database, and then click the New Query button.
3. Type the command or commands to save in the script. When necessary, click the Execute button to execute the commands saved in the script.
4. When you are finished, click the Save button, navigate to the location in which to save the script, and then enter a name for the script.

To view, edit, or run an existing script:

1. Load SQL Server Management Studio, and then click the Connect button in the Connect to Server dialog box.
2. Open the appropriate database, and then click the New Query button.
3. Click the Open File button on the toolbar.
4. Navigate to the folder containing the script file, select the script file, and then click the Open button in the Open File dialog box. (If the Connect to Database Engine dialog box opens, click the Connect button.) The script appears in the Query Editor window. You can view the contents of the script to make changes to it by editing the commands. If you edit a script, click the Save button to save your changes.
5. To run a script, click the Execute button.

CREATING THE REMAINING DATABASE TABLES

To create the remaining tables in the Premiere Products database, you need to execute the appropriate CREATE TABLE and INSERT commands. You should save these commands as scripts so you can re-create your database, if necessary, by running the scripts.

NOTE

Your instructor might give you the script files to use to create the tables in the Premiere Products, Henry Books, and Alexamara Marina Group databases and to insert data into them.

Figure 3-25 shows the CREATE TABLE command for the CUSTOMER table. Notice that the CUSTOMER_NAME column is specified as NOT NULL. Additionally, the CUSTOMER_NUM column is the table's primary key, indicating that the CUSTOMER_NUM column is the unique identifier of rows in the table. With this column designated as the primary key, the DBMS will reject any attempt to store a customer number that already exists in the table.

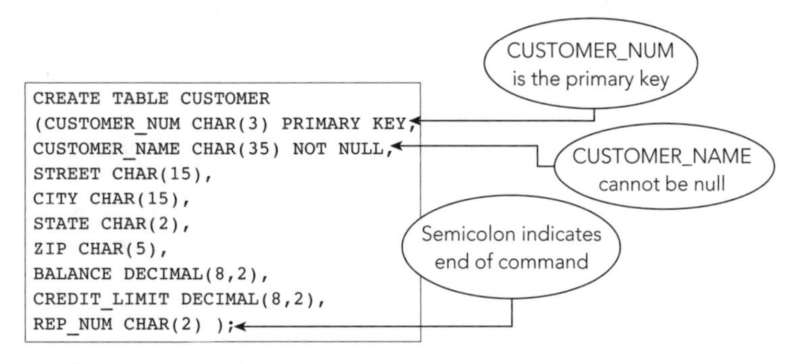

FIGURE 3-25 CREATE TABLE command for the CUSTOMER table

After creating the CUSTOMER table, you can create another script file containing the INSERT commands to add the customer rows to the table. When a script file contains more than one command, each command must end with a semicolon. Figure 3-26 shows the INSERT commands to add rows to the CUSTOMER table. As noted previously, to enter an apostrophe (single quotation mark) in the value for a field, type two single quotation marks, as illustrated in the name in the first INSERT command (Al's Appliance and Sport) in Figure 3-26.

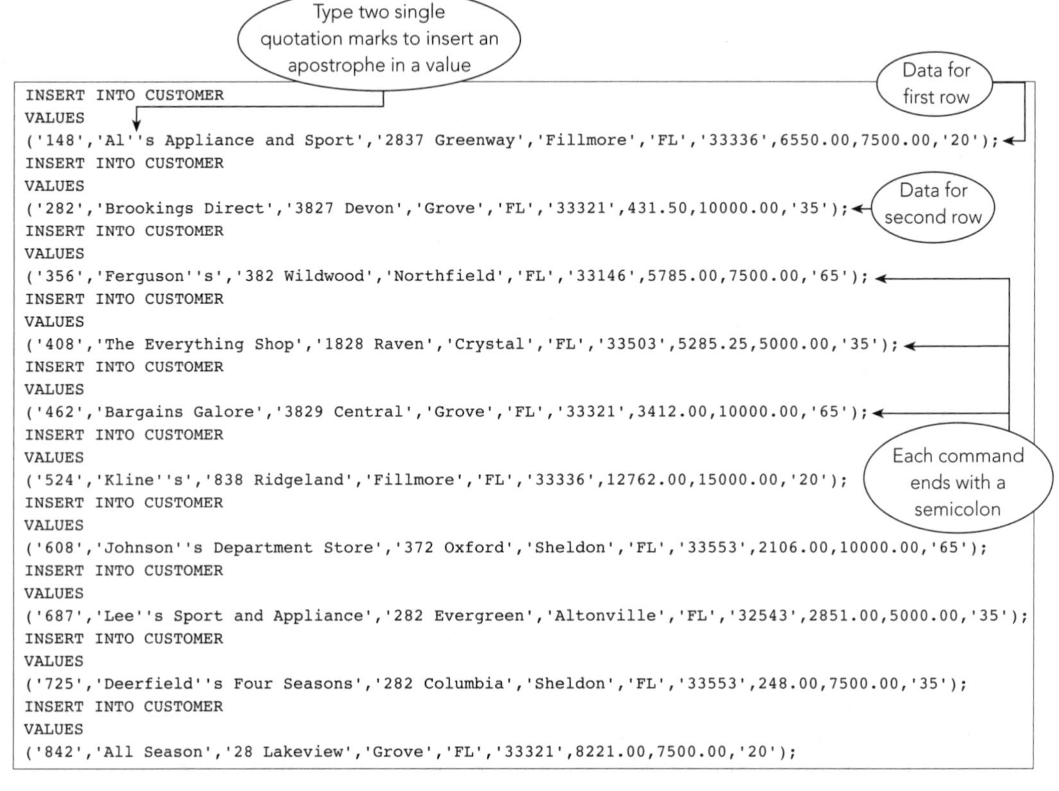

```
INSERT INTO CUSTOMER
VALUES
('148','Al''s Appliance and Sport','2837 Greenway','Fillmore','FL','33336',6550.00,7500.00,'20');
INSERT INTO CUSTOMER
VALUES
('282','Brookings Direct','3827 Devon','Grove','FL','33321',431.50,10000.00,'35');
INSERT INTO CUSTOMER
VALUES
('356','Ferguson''s','382 Wildwood','Northfield','FL','33146',5785.00,7500.00,'65');
INSERT INTO CUSTOMER
VALUES
('408','The Everything Shop','1828 Raven','Crystal','FL','33503',5285.25,5000.00,'35');
INSERT INTO CUSTOMER
VALUES
('462','Bargains Galore','3829 Central','Grove','FL','33321',3412.00,10000.00,'65');
INSERT INTO CUSTOMER
VALUES
('524','Kline''s','838 Ridgeland','Fillmore','FL','33336',12762.00,15000.00,'20');
INSERT INTO CUSTOMER
VALUES
('608','Johnson''s Department Store','372 Oxford','Sheldon','FL','33553',2106.00,10000.00,'65');
INSERT INTO CUSTOMER
VALUES
('687','Lee''s Sport and Appliance','282 Evergreen','Altonville','FL','32543',2851.00,5000.00,'35');
INSERT INTO CUSTOMER
VALUES
('725','Deerfield''s Four Seasons','282 Columbia','Sheldon','FL','33553',248.00,7500.00,'35');
INSERT INTO CUSTOMER
VALUES
('842','All Season','28 Lakeview','Grove','FL','33321',8221.00,7500.00,'20');
```


FIGURE 3-26 INSERT commands for the CUSTOMER table

Figures 3-27 through 3-32 show the scripts for the CREATE TABLE and INSERT commands for creating and inserting data into the ORDERS, PART, and ORDER_LINE tables in the Premiere Products database. Figure 3-27 contains the CREATE TABLE command for the ORDERS table.

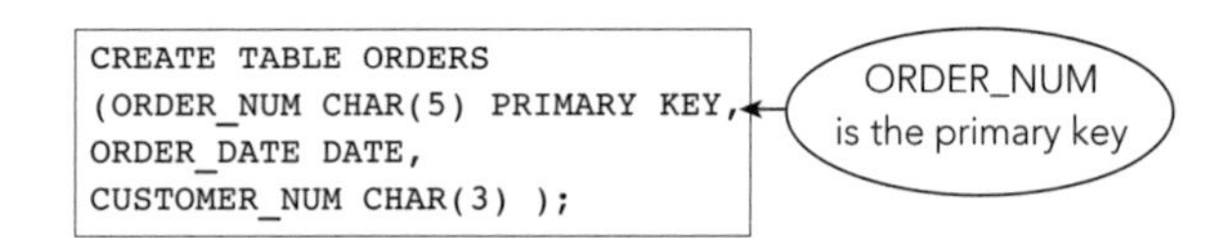

```
CREATE TABLE ORDERS
(ORDER_NUM CHAR(5) PRIMARY KEY,
ORDER_DATE DATE,
CUSTOMER_NUM CHAR(3) );
```


FIGURE 3-27 CREATE TABLE command for the ORDERS table

Figure 3-28 contains the INSERT commands to load data into the ORDERS table. Notice the way that dates are entered.

```
INSERT INTO ORDERS
VALUES
('21608','20-OCT-2010','148');
INSERT INTO ORDERS
VALUES
('21610','20-OCT-2010','356');
INSERT INTO ORDERS
VALUES
('21613','21-OCT-2010','408');
INSERT INTO ORDERS
VALUES
('21614','21-OCT-2010','282');
INSERT INTO ORDERS
VALUES
('21617','23-OCT-2010','608');
INSERT INTO ORDERS
VALUES
('21619','23-OCT-2010','148');
INSERT INTO ORDERS
VALUES
('21623','23-OCT-2010','608');
```


FIGURE 3-28 INSERT commands for the ORDERS table

Figure 3-29 contains the CREATE TABLE command for the PART table.

```
CREATE TABLE PART
(PART_NUM CHAR(4) PRIMARY KEY,
DESCRIPTION CHAR(15),
ON_HAND DECIMAL(4,0),
CLASS CHAR(2),
WAREHOUSE CHAR(1),
PRICE DECIMAL(6,2) );
```


FIGURE 3-29 CREATE TABLE command for the PART table

Figure 3-30 contains the INSERT commands to load data into the PART table.

```
INSERT INTO PART
VALUES
('AT94','Iron',50,'HW','3',24.95);
INSERT INTO PART
VALUES
('BV06','Home Gym',45,'SG','2',794.95);
INSERT INTO PART
VALUES
('CD52','Microwave Oven',32,'AP','1',165.00);
INSERT INTO PART
VALUES
('DL71','Cordless Drill',21,'HW','3',129.95);
INSERT INTO PART
VALUES
('DR93','Gas Range',8,'AP','2',495.00);
INSERT INTO PART
VALUES
('DW11','Washer',12,'AP','3',399.99);
INSERT INTO PART
VALUES
('FD21','Stand Mixer',22,'HW','3',159.95);
INSERT INTO PART
VALUES
('KL62','Dryer',12,'AP','1',349.95);
INSERT INTO PART
VALUES
('KT03','Dishwasher',8,'AP','3',595.00);
INSERT INTO PART
VALUES
('KV29','Treadmill',9,'SG','2',1390.00);
```

FIGURE 3-30 INSERT commands for the PART table

Figure 3-31 contains the CREATE TABLE command for the ORDER_LINE table. Notice the way that the primary key is defined when it consists of more than one column.

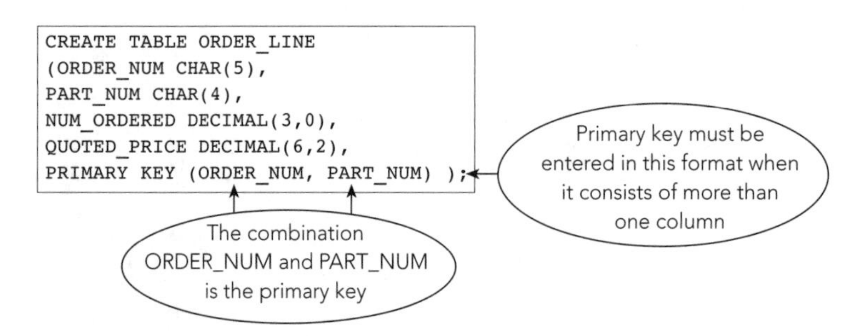

```
CREATE TABLE ORDER_LINE
(ORDER_NUM CHAR(5),
PART_NUM CHAR(4),
NUM_ORDERED DECIMAL(3,0),
QUOTED_PRICE DECIMAL(6,2),
PRIMARY KEY (ORDER_NUM, PART_NUM) );
```


FIGURE 3-31 CREATE TABLE command for the ORDER_LINE table

Figure 3-32 contains the INSERT commands to load data into the ORDER_LINE table.

```
INSERT INTO ORDER_LINE
VALUES
('21608','AT94',11,21.95);
INSERT INTO ORDER_LINE
VALUES
('21610','DR93',1,495.00);
INSERT INTO ORDER_LINE
VALUES
('21610','DW11',1,399.99);
INSERT INTO ORDER_LINE
VALUES
('21613','KL62',4,329.95);
INSERT INTO ORDER_LINE
VALUES
('21614','KT03',2,595.00);
INSERT INTO ORDER_LINE
VALUES
('21617','BV06',2,794.95);
INSERT INTO ORDER_LINE
VALUES
('21617','CD52',4,150.00);
INSERT INTO ORDER_LINE
VALUES
('21619','DR93',1,495.00);
INSERT INTO ORDER_LINE
VALUES
('21623','KV29',2,1290.00);
```

FIGURE 3-32 INSERT commands for the ORDER_LINE table

DESCRIBING A TABLE

The CREATE TABLE command defines a table's structure by listing its columns, data types, and column lengths. The CREATE TABLE command also indicates which columns cannot accept nulls. When you work with a table, you might not have access to the CREATE TABLE command that was used to create it. For example, another programmer might have created the table, or perhaps you created the table several months ago but did not save the command. You might want to examine the table's structure to see the details about the columns in the table. Each DBMS provides a method to examine a table's structure.

EXAMPLE 6

Describe the REP table.

In Oracle, you can use the **DESCRIBE** command to list all the columns in a table and their properties. Figure 3-33 shows the DESCRIBE command for the REP table. The result indicates the name of each column in the table, along with its data type and length. A value of 1 in the Primary Key column indicates the table's primary key column. A check mark in the Nullable column indicates a column that can accept null values. (The Precision, Scale, Default, and Comment columns in the results are beyond the scope of this discussion.)

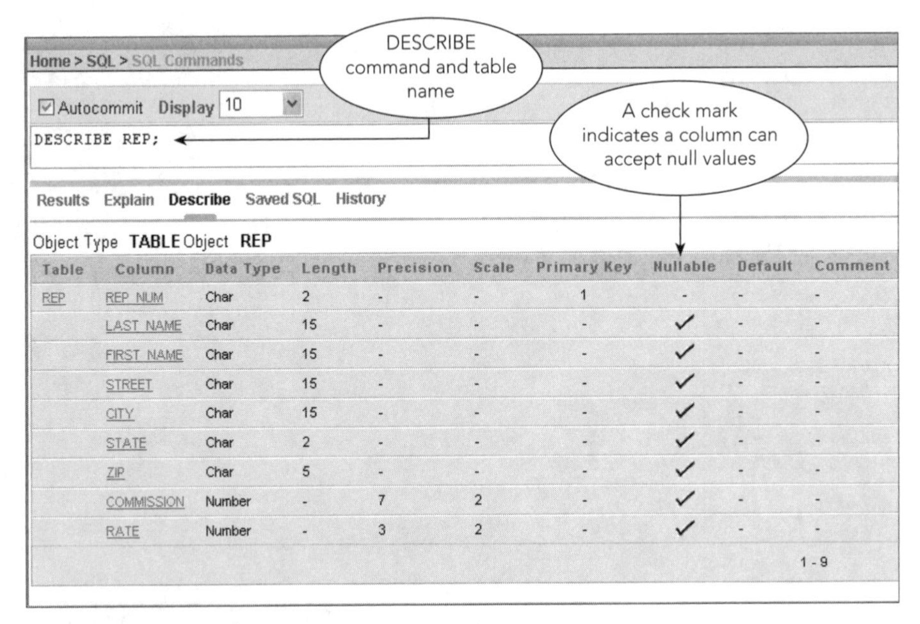

Table	Column	Data Type	Length	Precision	Scale	Primary Key	Nullable	Default	Comment
REP	REP_NUM	Char	2	-	-	1	-	-	-
	LAST_NAME	Char	15	-	-	-	✓	-	-
	FIRST_NAME	Char	15	-	-	-	✓	-	-
	STREET	Char	15	-	-	-	✓	-	-
	CITY	Char	15	-	-	-	✓	-	-
	STATE	Char	2	-	-	-	✓	-	-
	ZIP	Char	5	-	-	-	✓	-	-
	COMMISSION	Number	-	7	2	-	✓	-	-
	RATE	Number	-	3	2	-	✓	-	-

FIGURE 3-33 DESCRIBE command for the REP table

ACCESS USER NOTE

In Access, you use the Documenter tool to describe the tables (and other objects) in a database. To start the Documenter in Access 2003, open the database, click Tools on the menu bar, point to Analyze, and then click Documenter. To start the Documenter in Access 2007, click the Database Tools tab on the Ribbon, and then click the Database Documenter button in the Analyze group. In the Documenter dialog box, click the Tables tab, select the tables that you want to describe by putting a check mark in the check box next to their names, and then click the OK button. The Object Definition window opens and displays a report containing the requested documentation. You can customize the Documenter to control the amount of detail included in the report.

SQL SERVER USER NOTE

In SQL Server, you execute the sp_columns command to list all the columns in a table. The following command will list all the columns in the REP table:

```
Exec sp_columns REP
```

The result will indicate the name of each column in the REP table, along with its data type and length. A value of 1 in the Nullable column indicates a column that can accept null values. (The remaining columns that appear in the results are beyond the scope of this discussion.)

Chapter Summary

- Use the CREATE TABLE command to create a table by typing the table name and then listing within a single set of parentheses the columns in the table.
- Use the DROP TABLE command to delete a table and all its data from the database.
- Some commonly used data types in are INT, SMALLINT, DECIMAL, CHAR, VARCHAR, and DATE. Microsoft Access does not support DECIMAL. SQL Server uses DATETIME instead of DATE.
- A null data value (or null) is a special value that is used when the actual value for a column is unknown, unavailable, or not applicable.
- Use the NOT NULL clause in a CREATE TABLE command to identify columns that cannot accept null values.
- Use the INSERT command to insert rows into a table.
- Use the SELECT command to view the data in a table.
- Use the UPDATE command to change the value in a column.
- Use the DELETE command to delete a row from a table.
- You can save SQL commands in a script file in Oracle and SQL Server. In Microsoft Access, you save the commands as a query object in the database.
- You can use the DESCRIBE command in Oracle to display a table's structure and layout. In Access, you use the Documenter tool to produce a report of a table's structure and layout. In SQL Server, execute the sp_columns command to display the structure and layout of a table.

Key Terms

breadcrumb	null
CREATE TABLE	null data value
data type	script
DELETE	script file
DESCRIBE	script repository
DROP TABLE	SELECT
INSERT	Structured Query Language (SQL)
NOT NULL	UPDATE

Review Questions

1. How do you create a table using SQL?
2. How do you delete a table using SQL?
3. What are the common data types used to define columns using SQL?

4. Identify the best data type to use to store the following data in Oracle, in SQL Server, and in Access:
   a. The month, day, and year that an employee was hired
   b. An employee's Social Security number
   c. The department in which an employee works
   d. An employee's hourly pay rate

5. Write a paragraph that explains the difference between the CHAR data type and the VARCHAR data type. Use your Web browser and your favorite search engine to find examples of when to use VARCHAR and when to use CHAR. Be sure to cite the URL(s) that provided the examples as references at the end of your document.
6. What is a null value? How do you use SQL to identify columns that cannot accept null values?
7. Which SQL command do you use to add a row to a table?
8. Which SQL command do you use to view the data in a table?
9. Which SQL command do you use to change the value in a column in a table?
10. Which SQL command do you use to delete rows from a table?
11. How do you display the columns in a table and their characteristics in Oracle?

Exercises

To print a copy of your commands and results using Oracle, use the browser's Print command on the File menu or click the Print button on the browser's toolbar.

To print a copy of your commands and results using Access, use the instructions provided in the chapter to save your commands as query objects. To print a command, start Word or another word processor and create a new document. Select the SQL command in Access, copy it to the Clipboard, and then paste it into the document. To copy and paste a command's results, right-click the datasheet selector (the box in the upper-left corner of the datasheet) to select the entire datasheet, copy it to the Clipboard, and then paste it into the document.

To print a copy of your commands and results using SQL Server, start Word or another word processor and create a new document. Select the SQL command in SQL Server, copy it to the Clipboard, and then paste it into the document. To copy and paste a command's results, right-click the datasheet selector (the box in the upper-left corner of the datasheet) to select the entire datasheet, copy it to the Clipboard, and then paste it into the document.

Premiere Products

Use SQL to complete the following exercises.

1. Create a table named SALES_REP. The table has the same structure as the REP table shown in Figure 3-6 except the LAST_NAME column should use the VARCHAR data type and the COMMISSION and RATE columns should use the NUMBER data type. Execute the command to describe the layout and characteristics of the SALES_REP table.

2. Add the following row to the SALES_REP table: rep number: 25, last name: Lim; first name: Louis; street: 535 Vincent; city: Grove; state: FL; zip code: 33321; commission: 0.00; and rate: 0.05. Display the contents of the SALES_REP table.
3. Delete the SALES_REP table.
4. Run the script file for the Premiere database to create the five tables and add records to the tables. Be sure to select the file for the particular DBMS that you are using (Oracle, SQL Server, or Access). (*Note:* If you do not have the files for this text, ask your instructor for assistance.)
5. Confirm that you have created the tables correctly by describing each table and comparing the results to Figures 3-6, 3-25, 3-27, 3-29, and 3-31.
6. Confirm that you have added all data correctly by viewing the data in each table and comparing the results to Figure 2-1 in Chapter 2.

Henry Books

Use SQL to complete the following exercises.

1. Create a table named SALES_BRANCH. The table has the same structure as the BRANCH table shown in Figure 3-34 except the BRANCH_LOCATION column should use the VARCHAR data type and the BRANCH_NUM and NUM_EMPLOYEES columns should use the NUMBER data type. Execute the command to describe the layout and characteristics of the SALES_BRANCH table.
2. Add the following row to the SALES_BRANCH table: branch number: 5; branch name: Henry Town Plaza; branch location: 165 Plaza; and number of employees: 3. Display the contents of the SALES_BRANCH table.
3. Delete the SALES_BRANCH table.
4. Run the script file for the Henry Books database to create the six tables and add records to the tables. Be sure to select the file for the particular DBMS that you are using (Oracle, SQL Server, or Access). (*Note:* If you do not have the files for this text, ask your instructor for assistance.)
5. Confirm that you have created the tables correctly by describing each table and comparing the results to Figure 3-34.
6. Confirm that you have added all data correctly by viewing the data in each table and comparing the results to Figures 1-4 through 1-7 in Chapter 1.

BRANCH

Column	Type	Length	Decimal places	Nulls allowed?	Description
BRANCH_NUM	DECIMAL	2	0	No	Branch number (primary key)
BRANCH_NAME	CHAR	50			Branch name
BRANCH_LOCATION	CHAR	50			Branch location
NUM_EMPLOYEES	DECIMAL	2	0		Number of employees

PUBLISHER

Column	Type	Length	Decimal places	Nulls allowed?	Description
PUBLISHER_CODE	CHAR	3		No	Publisher code (primary key)
PUBLISHER_NAME	CHAR	25			Publisher name
CITY	CHAR	20			Publisher city

AUTHOR

Column	Type	Length	Decimal places	Nulls allowed?	Description
AUTHOR_NUM	DECIMAL	2	0	No	Author number (primary key)
AUTHOR_LAST	CHAR	12			Author last name
AUTHOR_FIRST	CHAR	10			Author first name

BOOK

Column	Type	Length	Decimal places	Nulls allowed?	Description
BOOK_CODE	CHAR	4		No	Book code (primary key)
TITLE	CHAR	40			Book title
PUBLISHER_CODE	CHAR	3			Publisher code
TYPE	CHAR	3			Book type
PRICE	DECIMAL	4	2		Book price
PAPERBACK	CHAR	1			Paperback (Y, N)

FIGURE 3-34 Table layouts for the Henry Books database

WROTE

Column	Type	Length	Decimal places	Nulls allowed?	Description
BOOK_CODE	CHAR	4		No	Book code (primary key)
AUTHOR_NUM	DECIMAL	2	0	No	Author number (primary key)
SEQUENCE	DECIMAL	1	0		Sequence number

INVENTORY

Column	Type	Length	Decimal places	Nulls allowed?	Description
BOOK_CODE	CHAR	4		No	Book code (primary key)
BRANCH_NUM	DECIMAL	2	0	No	Branch number (primary key)
ON_HAND	DECIMAL	2	0		Units on hand

FIGURE 3-34 Table layouts for the Henry Books database (continued)

Alexamara Marina Group

Use SQL to complete the following exercises.

1. Create a table named BOAT_SLIP. The table has the same structure as the MARINA_SLIP table shown in Figure 3-35 except the SLIP_ID, LENGTH, and RENTAL_FEE columns should use the NUMBER data type. Execute the command to describe the layout and characteristics of the BOAT_SLIP table.
2. Add the following record to the BOAT_SLIP table: slip ID: 12; marina number: 2; slip number: 7; length: 25; rental fee: 1800; boat name: Bavant; boat type: Ray 25; and owner number: FL13. Display the contents of the BOAT_SLIP table.
3. Delete the BOAT_SLIP table.
4. Run the script file for the Alexamara Marina Group database to create the five tables and add records to the tables. Be sure to select the file for the particular DBMS that you are using (Oracle, SQL Server, or Access). (*Note:* If you do not have the files for this text, ask your instructor for assistance.)
5. Confirm that you have created the tables correctly by describing each table and comparing the results to Figure 3-35.
6. Confirm that you have added all data correctly by viewing the data in each table and comparing the results to Figures 1-8 through 1-12 in Chapter 1.

MARINA

Column	Type	Length	Decimal places	Nulls allowed?	Description
MARINA_NUM	CHAR	4		No	Marina number (primary key)
NAME	CHAR	20			Marina name
ADDRESS	CHAR	15			Marina street address
CITY	CHAR	15			Marina city
STATE	CHAR	2			Marina state
ZIP	CHAR	5			Marina zip code

OWNER

Column	Type	Length	Decimal places	Nulls allowed?	Description
OWNER_NUM	CHAR	4		No	Owner number (primary key)
LAST_NAME	CHAR	50			Owner last name
FIRST_NAME	CHAR	20			Owner first name
ADDRESS	CHAR	15			Owner street address
CITY	CHAR	15			Owner city
STATE	CHAR	2			Owner state
ZIP	CHAR	5			Owner zip code

MARINA_SLIP

Column	Type	Length	Decimal places	Nulls allowed?	Description
SLIP_ID	DECIMAL	4	0	No	Slip ID (primary key)
MARINA_NUM	CHAR	4			Marina number
SLIP_NUM	CHAR	4			Slip number in the marina
LENGTH	DECIMAL	4	0		Length of slip (in feet)
RENTAL_FEE	DECIMAL	8	2		Annual rental fee for the slip
BOAT_NAME	CHAR	50			Name of boat currently in the slip
BOAT_TYPE	CHAR	50			Type of boat currently in the slip
OWNER_NUM	CHAR	4			Number of boat owner renting the slip

FIGURE 3-35 Table layouts for the Alexamara Marina Group database

SERVICE_CATEGORY

Column	Type	Length	Decimal places	Nulls allowed?	Description
CATEGORY_NUM	DECIMAL	4	0	No	Category number (primary key)
CATEGORY_DESCRIPTION	CHAR	255			Category description

SERVICE_REQUEST

Column	Type	Length	Decimal places	Nulls allowed?	Description
SERVICE_ID	DECIMAL	4	0	No	Service ID (primary key)
SLIP_ID	DECIMAL	4	0		Slip ID of the boat for which service is requested
CATEGORY_NUM	DECIMAL	4	0		Category number of the requested service
DESCRIPTION	CHAR	255			Description of specific service requested for boat
STATUS	CHAR	255			Description of status of service request
EST_HOURS	DECIMAL	4	2		Estimated number of hours required to complete the service
SPENT_HOURS	DECIMAL	4	2		Hours already spent on the service
NEXT_SERVICE_DATE	DATE				Next scheduled date for work on this service (or null if no next service date is specified)

FIGURE 3-35 Table layouts for the Alexamara Marina Group database (continued)

CHAPTER 4

SINGLE-TABLE QUERIES

LEARNING OBJECTIVES

Objectives

- Retrieve data from a database using SQL commands
- Use simple and compound conditions in queries
- Use the BETWEEN, LIKE, and IN operators in queries
- Use computed columns in queries
- Sort data using the ORDER BY clause
- Sort data using multiple keys and in ascending and descending order
- Use aggregate functions in a query
- Use subqueries
- Group data using the GROUP BY clause
- Select individual groups of data using the HAVING clause
- Retrieve columns with null values

INTRODUCTION

In this chapter, you will learn about the SQL SELECT command that is used to retrieve data in a database. You will examine ways to sort data and use SQL functions to count rows and calculate totals. You also will learn how to nest SELECT commands by placing one SELECT command inside another. Finally, you will learn how to group rows that have matching values in some column.

CONSTRUCTING SIMPLE QUERIES

One of the most important features of a DBMS is its ability to answer a wide variety of questions concerning the data in a database. When you need to find data that answers a specific question, you use a query. A **query** is a question represented in a way that the DBMS can understand.

In SQL, you use the SELECT command to query a database. The basic form of the SELECT command is SELECT-FROM-WHERE. After you type the word SELECT, you list the columns that you want to include in the query results. This portion of the command is called the **SELECT clause**. Next, you type the word FROM followed by the name of the table that contains the data you need to query. This portion of the command is called the **FROM clause**. Finally, after the word WHERE, you list any conditions (restrictions) that apply to the data you want to retrieve. This optional portion of the command is called the **WHERE clause**. For example, when you need to retrieve the rows for only those customers with credit limits of $7,500, include a condition in the WHERE clause specifying that the value in the CREDIT_LIMIT column must be $7,500 (CREDIT_LIMIT = 7500).

There are no special formatting rules in SQL. In this text, the FROM clause and the WHERE clause (when it is used) appear on separate lines only to make the commands more readable and understandable.

Retrieving Certain Columns and All Rows

You can write a command to retrieve specified columns and all rows from a table, as illustrated in Example 1.

EXAMPLE 1

List the number, name, and balance for all customers.

Because you need to list *all* customers, you do not need to include a WHERE clause; you do not need to put any restrictions on the data to retrieve. You simply list the columns to be included (CUSTOMER_NUM, CUSTOMER_NAME, and BALANCE) in the SELECT clause and the name of the table (CUSTOMER) in the FROM clause. Type a semicolon to indicate the end of the command, and then click the Run button to display the results. The query and its results appear in Figure 4-1.

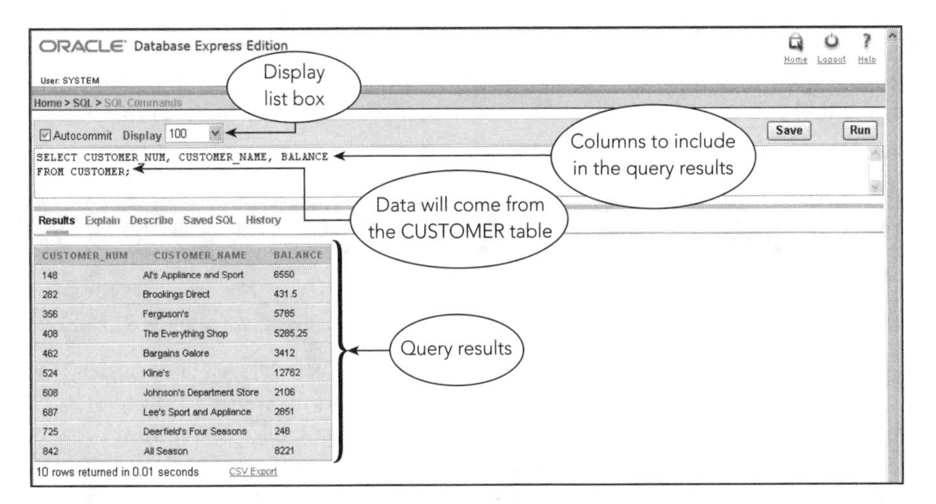

FIGURE 4-1 SELECT command to select certain columns from the CUSTOMER table

NOTE

In the Oracle Database Express Edition, the number in the Display list box indicates the maximum number of rows that Oracle will display in the query results. The default value is 10. To change the value, either click the arrow and select a new value from the list or type a new value in the box. Figure 4-1 shows the Display list box after the user changed it to display 100 rows. When you run a query whose results will include more rows than the number in the Display list box, Oracle will display a message indicating this fact. If this situation occurs, increase the number in the Display list box, and then click the Run button again to display the complete query results.

NOTE

If you are using Access or SQL Server to run the SQL commands shown in this text, your query results will differ slightly from the results shown in the figures. In Access, the BALANCE field has the CURRENCY data type and Access will display values in this column with two decimal places and a dollar sign. In SQL Server, values in the BALANCE field will be displayed with two decimal places and DATE field values might be displayed with a time value. Although your output might be formatted differently, the data should be the same as what you see in the figures.

Retrieving All Columns and All Rows

You can use the same type of command illustrated in Example 1 to retrieve all columns and all rows from a table. As Example 2 illustrates, however, you can use a shortcut to accomplish this task.

EXAMPLE 2

List the complete PART table.

Instead of including every column in the SELECT clause, you can use an asterisk (*) to indicate that you want to include all columns. The result lists all columns in the order in which you described them to the DBMS when you created the table. If you want the columns listed in a different order, type the column names in the order in which you want them to appear in the query results. In this case, assuming that the default order is appropriate, you can use the query shown in Figure 4-2 to display the complete PART table.

```
SELECT *
FROM PART;
```

Asterisk indicates all columns will be included

Results Explain Describe Saved SQL History

PART_NUM	DESCRIPTION	ON_HAND	CLASS	WAREHOUSE	PRICE
AT94	Iron	50	HW	3	24.95
BV06	Home Gym	45	SG	2	794.95
CD52	Microwave Oven	32	AP	1	165
DL71	Cordless Drill	21	HW	3	129.95
DR93	Gas Range	8	AP	2	495
DW11	Washer	12	AP	3	399.99
FD21	Stand Mixer	22	HW	3	159.95
KL62	Dryer	12	AP	1	349.95
KT03	Dishwasher	8	AP	3	595
KV29	Treadmill	9	SG	2	1390

10 rows returned in 0.04 seconds CSV Export

FIGURE 4-2 SELECT command to select all columns from the PART table

Using a WHERE Clause

When you need to retrieve rows that satisfy some condition, you include a WHERE clause in the SELECT command, as shown in Example 3.

EXAMPLE 3

What is the name of the customer with customer number 148?

You can use a WHERE clause to restrict the query results to customer number 148, as shown in Figure 4-3. Because CUSTOMER_NUM is a character column, the value 148 is enclosed in single quotation marks. In addition, because the CUSTOMER_NUM column is the primary key of the CUSTOMER table, there can be only one customer whose number matches the number in the WHERE clause.

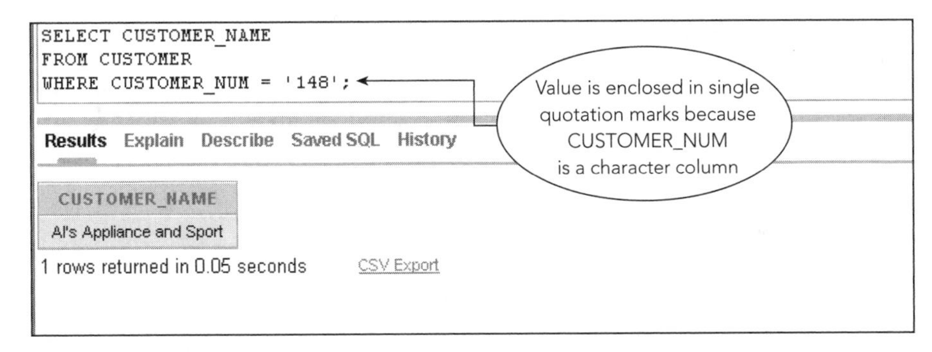

FIGURE 4-3 SELECT command to find the name of customer number 148

The condition in the preceding WHERE clause is called a simple condition. A **simple condition** has the form column name, comparison operator, and then either another column name or a value. Figure 4-4 lists the comparison operators that you can use in SQL. Notice that there are two versions of the "not equal to" operator: < > and !=.

Comparison operator	Description
=	Equal to
<	Less than
>	Greater than
<=	Less than or equal to
>=	Greater than or equal to
< >	Not equal to
!=	Not equal to

FIGURE 4-4 Comparison operators used in SQL commands

EXAMPLE 4

Find the number and name of each customer located in the city of Grove.

The only difference between this example and the previous one is that in Example 3, there could not be more than one row in the answer because the condition involved the table's primary key. In Example 4, the condition involves a column that is *not* the table's primary key. Because there is more than one customer located in the city of Grove, the results can and do contain more than one row, as shown in Figure 4-5.

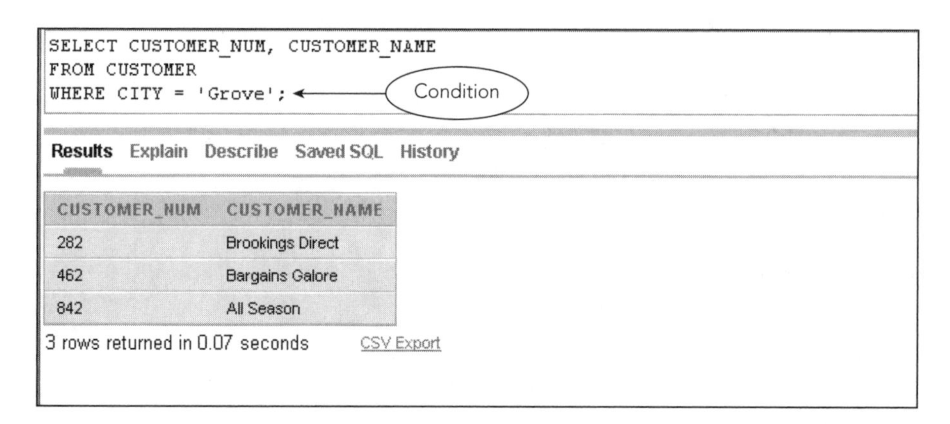

FIGURE 4-5 SELECT command to find all customers located in Grove

EXAMPLE 5

Find the number, name, balance, and credit limit for all customers with balances that exceed their credit limits.

A simple condition can also compare the values stored in two columns. In Figure 4-6, the WHERE clause includes a comparison operator that selects only those rows in which the balance is greater than the credit limit.

```
SELECT CUSTOMER_NUM, CUSTOMER_NAME, BALANCE, CREDIT_LIMIT
FROM CUSTOMER
WHERE BALANCE > CREDIT_LIMIT;
```

Results Explain Describe Saved SQL History

CUSTOMER_NUM	CUSTOMER_NAME	BALANCE	CREDIT_LIMIT
408	The Everything Shop	5285.25	5000
842	All Season	8221	7500

2 rows returned in 0.03 seconds CSV Export

FIGURE 4-6 SELECT command to find all customers with balances that exceed their credit limits

Using Compound Conditions

The conditions you have seen so far are called simple conditions. The following examples require compound conditions. You form a **compound condition** by connecting two or more simple conditions with the AND, OR, and NOT operators. When the **AND** operator connects simple conditions, all the simple conditions must be true in order for the compound condition to be true. When the **OR** operator connects the simple conditions, the compound condition will be true whenever any one of the simple conditions is true. Preceding a condition by the **NOT** operator reverses the truth of the original condition. For example, if the original condition is true, the new condition will be false; if the original condition is false, the new one will be true.

EXAMPLE 6

List the descriptions of all parts that are located in warehouse 3 and for which there are more than 25 units on hand.

In Example 6, you need to retrieve those parts that meet *both* conditions—the warehouse number is equal to 3 *and* the number of units on hand is greater than 25. To find the answer, you form a compound condition using the AND operator, as shown in Figure 4-7. The query examines the data in the PART table and lists the parts that are located in warehouse 3 and for which there are more than 25 units on hand. When a WHERE clause uses the AND operator to connect simple conditions, it also is called an **AND condition**.

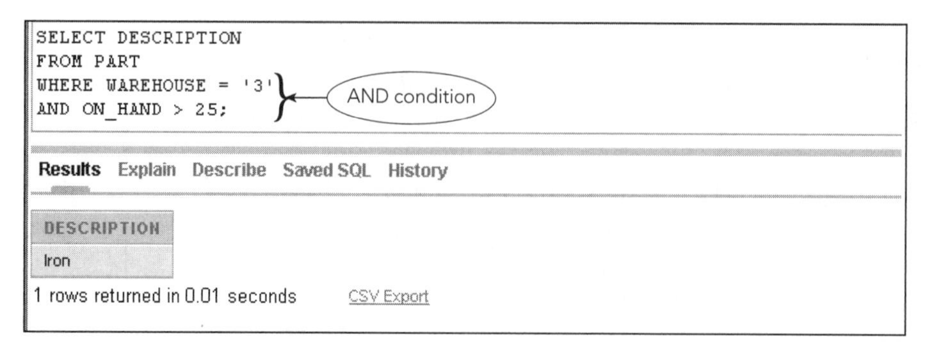

FIGURE 4-7 SELECT command with an AND condition on separate lines

For readability, each of the simple conditions in the query shown in Figure 4-7 appears on a separate line. Some people prefer to put the conditions on the same line with parentheses around each simple condition, as shown in Figure 4-8. These two methods accomplish the same thing. In this text, simple conditions will appear on separate lines and without parentheses.

```
SELECT DESCRIPTION
FROM PART
WHERE (WAREHOUSE = '3') AND (ON_HAND > 25);
```

Results Explain Describe Saved SQL History

DESCRIPTION
Iron

1 rows returned in 0.01 seconds CSV Export

FIGURE 4-8 SELECT command with an AND condition on a single line

EXAMPLE 7

List the descriptions of all parts that are located in warehouse 3 or for which there are more than 25 units on hand.

In Example 7, you need to retrieve descriptions for those parts for which the warehouse number is equal to 3, *or* the number of units on hand is greater than 25, *or* both. To do this, you form a compound condition using the OR operator, as shown in Figure 4-9. When a WHERE clause uses the OR operator to connect simple conditions, it also is called an **OR condition**.

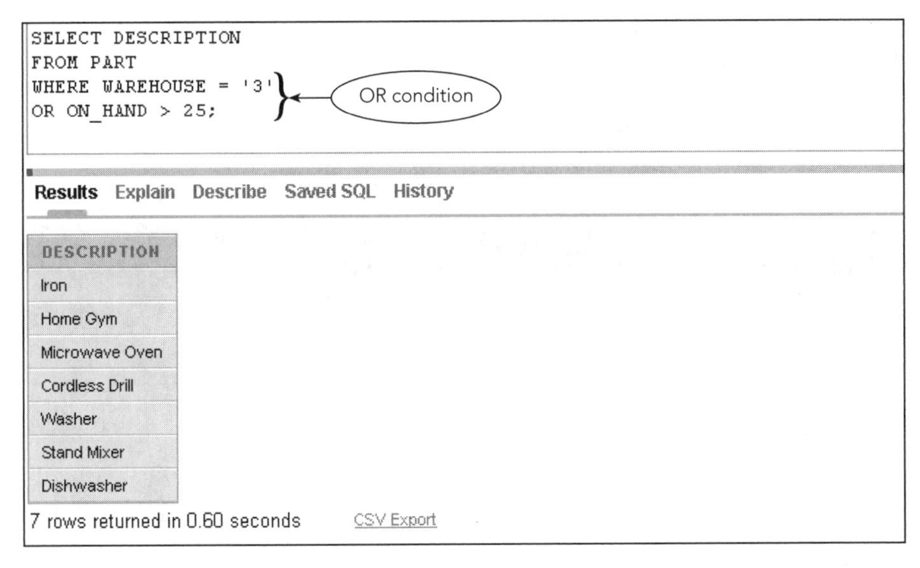
```
SELECT DESCRIPTION
FROM PART
WHERE WAREHOUSE = '3'
OR ON_HAND > 25;
```


FIGURE 4-9 SELECT command with an OR condition

EXAMPLE 8

List the descriptions of all parts that are not in warehouse 3.

For Example 8, you could use a simple condition and the "not equal to" operator (WHERE WAREHOUSE < > '3'). As an alternative, you could use the EQUAL operator (=) in the condition and precede the entire condition with the NOT operator, as shown in Figure 4-10. When a WHERE clause uses the NOT operator to connect simple conditions, it also is called a **NOT condition**.

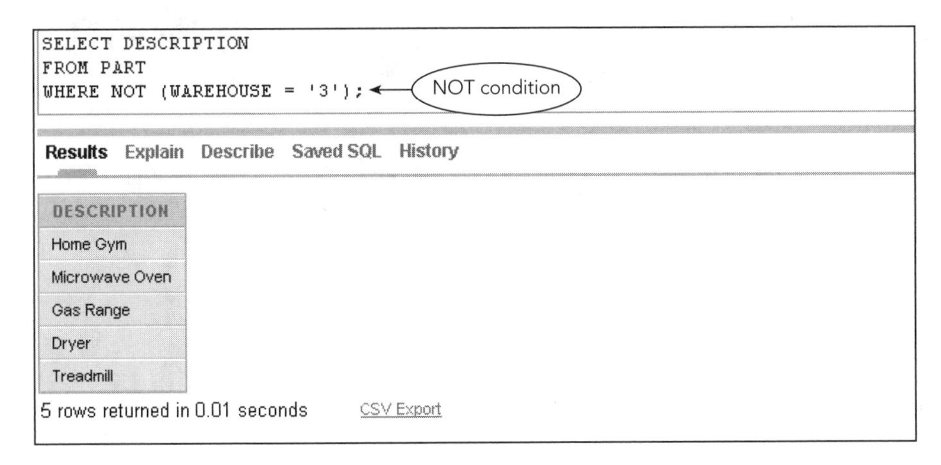
```
SELECT DESCRIPTION
FROM PART
WHERE NOT (WAREHOUSE = '3');
```


FIGURE 4-10 SELECT command with a NOT condition

You do not need to enclose the condition WAREHOUSE = '3' in parentheses, but doing so makes the command more readable.

Using the BETWEEN Operator

Example 9 requires a compound condition to determine the answer.

EXAMPLE 9

List the number, name, and balance of all customers with balances greater than or equal to $2,000 and less than or equal to $5,000.

You can use a WHERE clause and the AND operator, as shown in Figure 4-11, to retrieve the data.

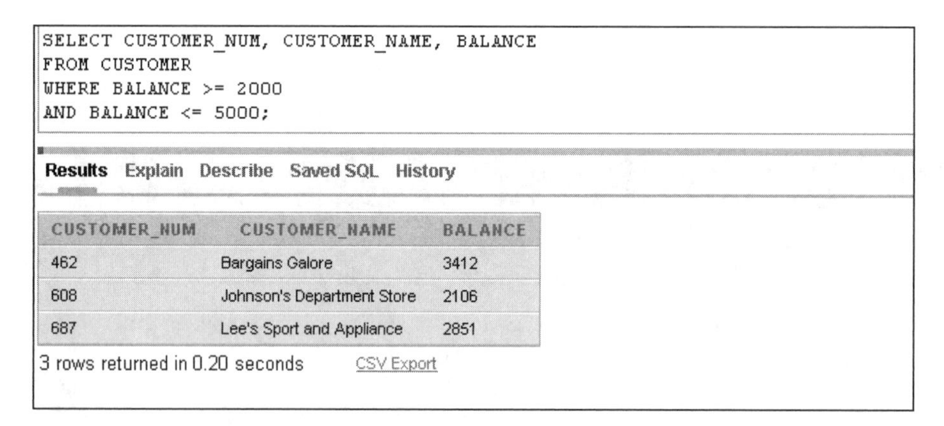

```
SELECT CUSTOMER_NUM, CUSTOMER_NAME, BALANCE
FROM CUSTOMER
WHERE BALANCE >= 2000
AND BALANCE <= 5000;
```

Results Explain Describe Saved SQL History

CUSTOMER_NUM	CUSTOMER_NAME	BALANCE
462	Bargains Galore	3412
608	Johnson's Department Store	2106
687	Lee's Sport and Appliance	2851

3 rows returned in 0.20 seconds CSV Export

FIGURE 4-11 SELECT command with an AND condition for a single column

NOTE

In SQL, numbers included in queries are entered without extra symbols, such as dollar signs and commas.

An alternative to this approach uses the BETWEEN operator, as shown in Figure 4-12. The **BETWEEN** operator lets you specify a range of values in a condition.

```
SELECT CUSTOMER_NUM, CUSTOMER_NAME, BALANCE
FROM CUSTOMER
WHERE BALANCE BETWEEN 2000 AND 5000;
```

Results Explain Describe Saved SQL History

CUSTOMER_NUM	CUSTOMER_NAME	BALANCE
462	Bargains Galore	3412
608	Johnson's Department Store	2106
687	Lee's Sport and Appliance	2851

3 rows returned in 0.00 seconds CSV Export

FIGURE 4-12 SELECT command with the BETWEEN operator

The BETWEEN operator is inclusive, meaning that the query selects a value equal to either value in the condition and in the range of the values. In the clause BETWEEN 2000 and 5000, for example, values of 2,000 through 5,000 would make the condition true. You can use the BETWEEN operator in Oracle, SQL Server, and Access.

The BETWEEN operator is not an essential feature of SQL; you have just seen that you can obtain the same result without it. Using the BETWEEN operator, however, does make certain SELECT commands simpler to construct.

Using Computed Columns

You can perform computations using SQL queries. A **computed column** does not exist in the database but can be computed using data in the existing columns. Computations can involve any arithmetic operator shown in Figure 4-13.

Arithmetic operator	Description
+	Addition
–	Subtraction
*	Multiplication
/	Division

FIGURE 4-13 Arithmetic operators

EXAMPLE 10

Find the number, name, and available credit (the credit limit minus the balance) for each customer.

There is no column in the Premiere Products database that stores a customer's available credit, but you can compute the available credit using the CREDIT_LIMIT and BALANCE columns. To compute the available credit, you use the expression CREDIT_LIMIT - BALANCE, as shown in Figure 4-14.

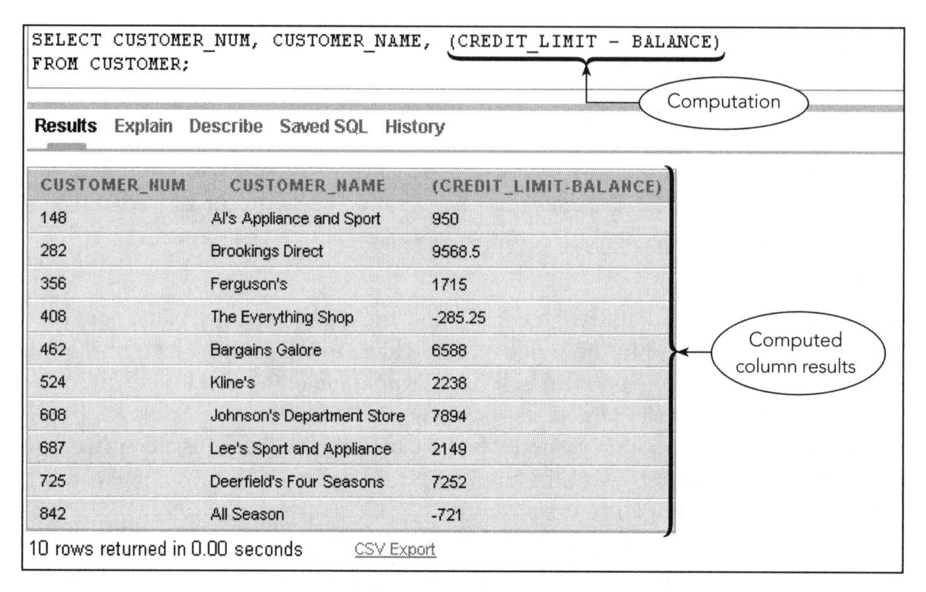
SELECT CUSTOMER_NUM, CUSTOMER_NAME, (CREDIT_LIMIT - BALANCE)
FROM CUSTOMER;

CUSTOMER_NUM	CUSTOMER_NAME	(CREDIT_LIMIT-BALANCE)
148	Al's Appliance and Sport	950
282	Brookings Direct	9568.5
356	Ferguson's	1715
408	The Everything Shop	-285.25
462	Bargains Galore	6588
524	Kline's	2238
608	Johnson's Department Store	7894
687	Lee's Sport and Appliance	2149
725	Deerfield's Four Seasons	7252
842	All Season	-721

FIGURE 4-14 SELECT command with a computed column

The parentheses around the calculation (CREDIT_LIMIT - BALANCE) are not essential but improve readability.

You also can assign a name to a computed column by following the computation with the word AS and the desired name. The command shown in Figure 4-15, for example, assigns the name AVAILABLE_CREDIT to the computed column.

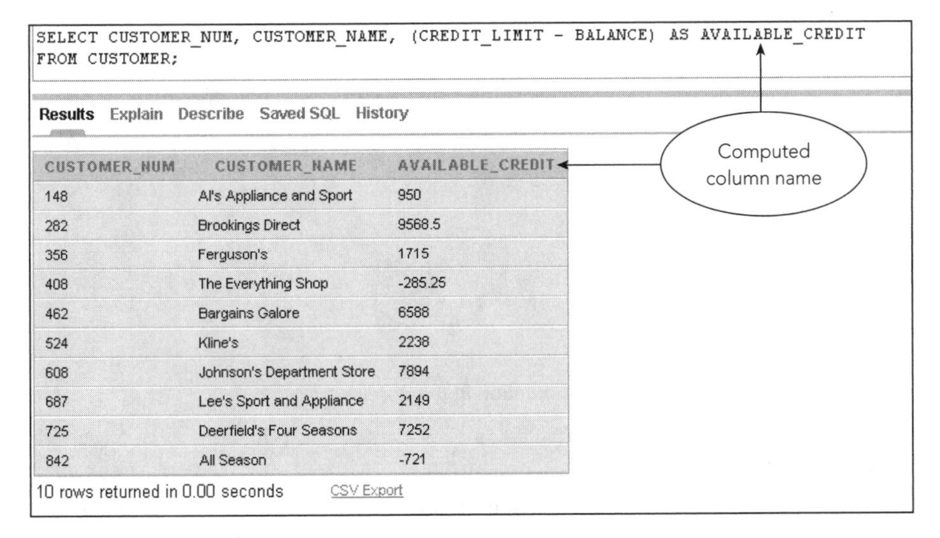
SELECT CUSTOMER_NUM, CUSTOMER_NAME, (CREDIT_LIMIT - BALANCE) AS AVAILABLE_CREDIT
FROM CUSTOMER;

Results Explain Describe Saved SQL History

CUSTOMER_NUM	CUSTOMER_NAME	AVAILABLE_CREDIT
148	Al's Appliance and Sport	950
282	Brookings Direct	9568.5
356	Ferguson's	1715
408	The Everything Shop	-285.25
462	Bargains Galore	6588
524	Kline's	2238
608	Johnson's Department Store	7894
687	Lee's Sport and Appliance	2149
725	Deerfield's Four Seasons	7252
842	All Season	-721

10 rows returned in 0.00 seconds CSV Export

FIGURE 4-15 SELECT command with a named computed column

NOTE

You can use names containing spaces following the word AS. In many SQL implementations, including Oracle, you do so by enclosing the name in quotation marks (for example, AS "AVAILABLE CREDIT"). Other SQL implementations require you to enclose the name in other special characters. For example, in Access you would enclose the name in square brackets (AS [AVAILABLE CREDIT]). In SQL Server, you can use either quotation marks or square brackets.

EXAMPLE 11

Find the number, name, and available credit for each customer with at least $5,000 of available credit.

You also can use computed columns in comparisons, as shown in Figure 4-16.

```
SELECT CUSTOMER_NUM, CUSTOMER_NAME, (CREDIT_LIMIT - BALANCE) AS AVAILABLE_CREDIT
FROM CUSTOMER
WHERE (CREDIT_LIMIT - BALANCE) >= 5000;
```

Results Explain Describe Saved SQL History

CUSTOMER_NUM	CUSTOMER_NAME	AVAILABLE_CREDIT
282	Brookings Direct	9568.5
462	Bargains Galore	6588
608	Johnson's Department Store	7894
725	Deerfield's Four Seasons	7252

4 rows returned in 0.02 seconds CSV Export

FIGURE 4-16 SELECT command with a computation in the condition

Using the LIKE Operator

In most cases, the conditions in WHERE clauses involve exact matches, such as retrieving rows for each customer located in the city of Grove. In some cases, however, exact matches do not work. For example, you might know that the desired value contains only a certain collection of characters. In such cases, you use the LIKE operator with a wildcard symbol, as shown in Example 12. Rather than testing for equality, the **LIKE** operator uses one or more wildcard characters to test for a pattern match.

EXAMPLE 12

List the number, name, and complete address of each customer located on a street that contains the letters "Central."

All you know is that the addresses you want contain a certain collection of characters ("Central") somewhere in the STREET column, but you do not know where. In SQL for Oracle and for SQL Server, the percent sign (%) is used as a wildcard to represent any collection of characters. As shown in Figure 4-17, the condition LIKE '%Central%' retrieves information for each customer whose street contains some collection of characters, followed by the letters "Central," followed potentially by some additional characters. Note that this query also would retrieve information for a customer whose address is "123 Centralia" because "Centralia" also contains the letters "Central."

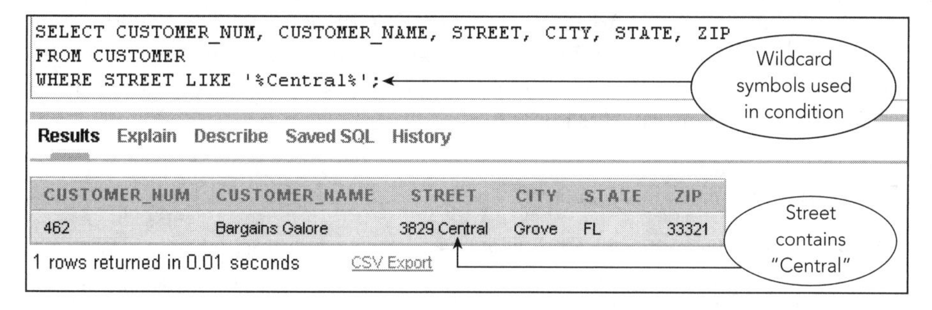

FIGURE 4-17 SELECT command with a LIKE operator and wildcards

Another wildcard symbol in SQL is the underscore (_), which represents any individual character. For example, "T_m" represents the letter "T" followed by any single character, followed by the letter "m," and would retrieve rows that include the words Tim, Tom, or T3m.

ACCESS USER NOTE

Access uses different wildcard symbols. The symbol for any collection of characters is the asterisk (*), as shown in Figure 4-18. The symbol for an individual character is the question mark (?).

```
SELECT CUSTOMER_NUM, CUSTOMER_NAME, STREET, CITY, STATE, ZIP
FROM CUSTOMER
WHERE STREET LIKE '*Central*';
```

FIGURE 4-18 Access SELECT command with wildcards

NOTE

In a large database, you should use wildcards only when absolutely necessary. Searches involving wildcards can be extremely slow to process.

Using the IN Operator

An **IN clause**, which consists of the IN operator followed by a collection of values, provides a concise way of phrasing certain conditions, as Example 13 illustrates. You will see another use for the IN clause in more complex examples later in this chapter.

EXAMPLE 13

List the number, name, and credit limit for each customer with a credit limit of $5,000, $10,000, or $15,000.

In this query, you can use an IN clause to determine whether a credit limit is $5,000, $10,000, or $15,000. You could obtain the same answer by using the condition WHERE

CREDIT_LIMIT = 5000 OR CREDIT_LIMIT = 10000 OR CREDIT_LIMIT = 15000. The approach shown in Figure 4-19 is simpler because the IN clause contains a collection of values: 5000, 10000, and 15000. The condition is true for those rows in which the value in the CREDIT_LIMIT column is in this collection.


```
SELECT CUSTOMER_NUM, CUSTOMER_NAME, CREDIT_LIMIT
FROM CUSTOMER
WHERE CREDIT_LIMIT IN (5000, 10000, 15000);
```

List of values in the IN clause

Results Explain Describe Saved SQL History

CUSTOMER_NUM	CUSTOMER_NAME	CREDIT_LIMIT
282	Brookings Direct	10000
408	The Everything Shop	5000
462	Bargains Galore	10000
524	Kline's	15000
608	Johnson's Department Store	10000
687	Lee's Sport and Appliance	5000

6 rows returned in 0.01 seconds CSV Export

FIGURE 4-19 SELECT command with an IN clause

SORTING

Recall that the order of rows in a table is immaterial to the DBMS. From a practical standpoint, this means that when you query a relational database, there is no defined order in which to display the results. Rows might be displayed in the order in which the data was originally entered, but even this is not certain. If the order in which the data is displayed is important, you can specifically request that the results appear in a desired order. In SQL, you specify the results order by using the ORDER BY clause.

Using the ORDER BY Clause

You use the **ORDER BY clause** to list data in a specific order, as shown in Example 14.

EXAMPLE 14

List the number, name, and balance of each customer. Order (sort) the output in ascending (increasing) order by balance.

The column on which to sort data is called a **sort key** or simply a **key**. In Example 14, you need to order the output by balance, so the sort key is the BALANCE column. To sort the output, use an ORDER BY clause followed by the sort key. If you do not specify a sort order, the default is ascending. The query appears in Figure 4-20.

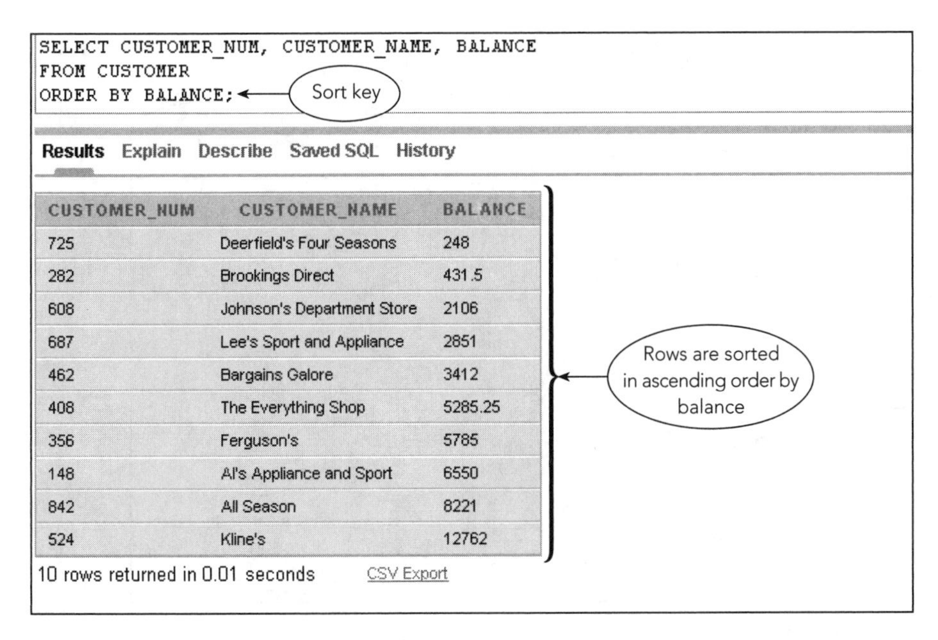

CUSTOMER_NUM	CUSTOMER_NAME	BALANCE
725	Deerfield's Four Seasons	248
282	Brookings Direct	431.5
608	Johnson's Department Store	2106
687	Lee's Sport and Appliance	2851
462	Bargains Galore	3412
408	The Everything Shop	5285.25
356	Ferguson's	5785
148	Al's Appliance and Sport	6550
842	All Season	8221
524	Kline's	12762

FIGURE 4-20 SELECT command to sort rows

Additional Sorting Options

Sometimes you might need to sort data using more than one key, as shown in Example 15.

EXAMPLE 15

List the number, name, and credit limit of each customer. Order the customers by name within descending credit limit. (In other words, first sort the customers by credit limit in descending order. Within each group of customers that have a common credit limit, sort the customers by name in ascending order.)

Example 15 involves two new ideas: sorting on multiple keys—CREDIT_LIMIT and CUSTOMER_NAME—and sorting one of the keys in descending order. When you need to sort data on two columns, the more important column (in this case, CREDIT_LIMIT) is called the **major sort key** (or the **primary sort key**) and the less important column (in this case, CUSTOMER_NAME) is called the **minor sort key** (or the **secondary sort key**). To sort on multiple keys, you list the keys in order of importance in the ORDER BY clause. To sort in descending order, you follow the name of the sort key with the **DESC** operator, as shown in Figure 4-21.

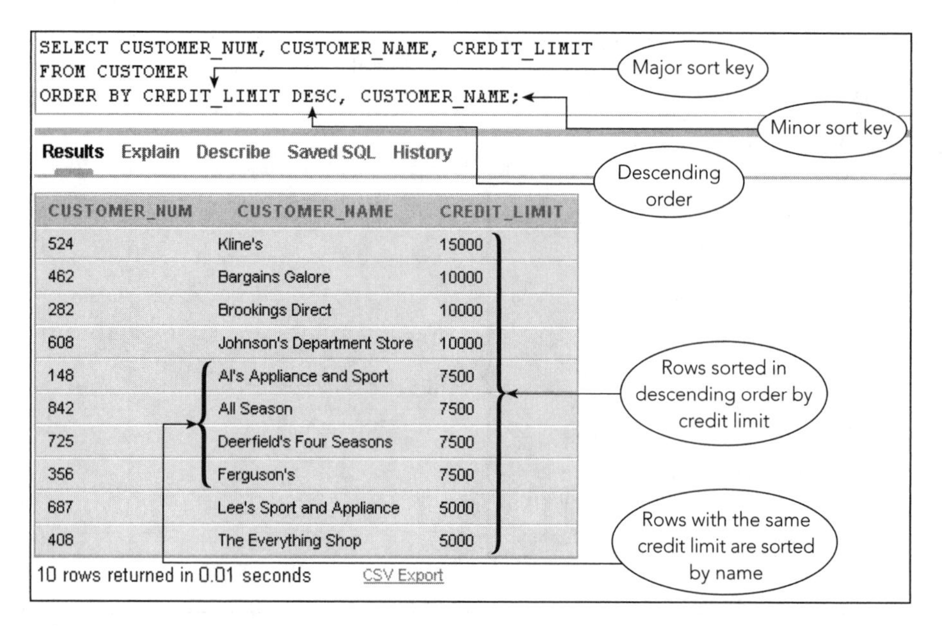

FIGURE 4-21 SELECT command to sort data using multiple sort keys

USING FUNCTIONS

SQL uses special functions, called **aggregate functions**, to calculate sums, averages, counts, maximum values, and minimum values. These functions apply to *groups* of rows. They could apply to all the rows in a table (for example, calculating the average balance of all customers). They also could apply to those rows satisfying some particular condition (for example, the average balance of all customers of sales rep 20). The descriptions of the aggregate functions appear in Figure 4-22.

Function	Description
AVG	Calculates the average value in a column
COUNT	Determines the number of rows in a table
MAX	Determines the maximum value in a column
MIN	Determines the minimum value in a column
SUM	Calculates a total of the values in a column

FIGURE 4-22 SQL aggregate functions

Using the COUNT Function

The **COUNT** function, as illustrated in Example 16, counts the number of rows in a table.

EXAMPLE 16

How many parts are in item class HW?

For this query, you need to determine the total number of rows in the PART table with the value HW in the CLASS column. You could count the part numbers in the query results, or the number of part descriptions, or the number of entries in any other column. It doesn't matter which column you choose because all columns should provide the same answer. Rather than arbitrarily selecting one column, most SQL implementations let you use the asterisk (*) to represent any column, as shown in Figure 4-23.

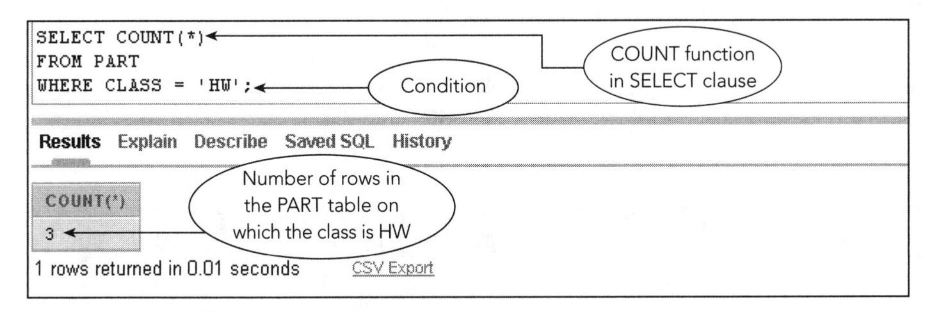

FIGURE 4-23 SELECT command to count rows

You also can count the number of rows in a query by selecting a specific column instead of using the asterisk, as follows:

```
SELECT COUNT(PART_NUM)
FROM PART
WHERE CLASS = 'HW';
```

Using the SUM Function

If you need to calculate the total of all customers' balances, you can use the **SUM** function, as illustrated in Example 17.

EXAMPLE 17

Find the total number of Premiere Products customers and the total of their balances.

When you use the SUM function, you must specify the column to total, and the column's data type must be numeric. (How could you calculate a sum of names or addresses?) Figure 4-24 shows the query.

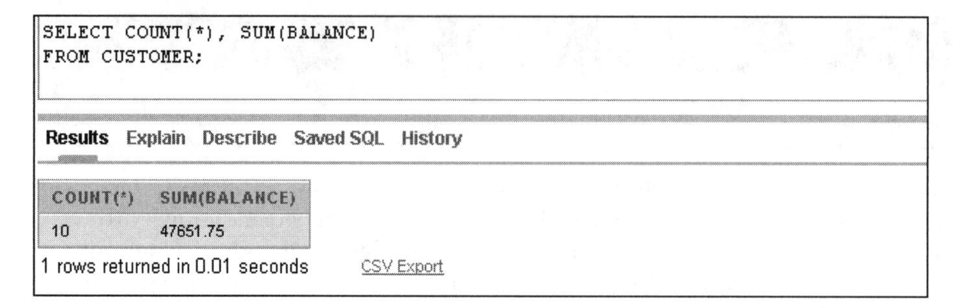

```
SELECT COUNT(*), SUM(BALANCE)
FROM CUSTOMER;
```


COUNT(*)	SUM(BALANCE)
10	47651.75

FIGURE 4-24 SELECT command to count rows and calculate a total

Using the AVG, MAX, and MIN Functions

Using the AVG, MAX, and MIN functions is similar to using SUM, except that different statistics are calculated. **AVG** calculates the average value in a numeric range, **MAX** calculates the maximum value in a numeric range, and **MIN** calculates the minimum value in a numeric range.

EXAMPLE 18

Find the sum of all balances, the average balance, the maximum balance, and the minimum balance of all Premiere Products customers.

Figure 4-25 shows the query and the results.

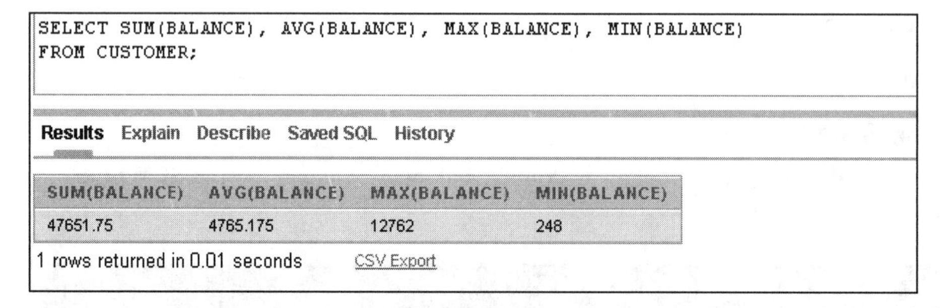

```
SELECT SUM(BALANCE), AVG(BALANCE), MAX(BALANCE), MIN(BALANCE)
FROM CUSTOMER;
```


SUM(BALANCE)	AVG(BALANCE)	MAX(BALANCE)	MIN(BALANCE)
47651.75	4765.175	12762	248

FIGURE 4-25 SELECT command with several functions

NOTE

When you use the SUM, AVG, MAX, or MIN functions, SQL ignores any null value(s) in the column and eliminates them from the computations.

Null values in numeric columns can produce strange results when statistics are computed. Suppose the BALANCE column accepts null values, there are currently four customers in the CUSTOMER table, and their respective balances are $100, $200, $300, and null (unknown). When you calculate the average balance, SQL ignores the null value and obtains a result of $200 (($100 + $200 + $300) / 3). Similarly, when you calculate the total of the balances, SQL ignores the null value and calculates a total of $600. When you count the number of customers in the table, however, SQL includes the row containing the null value, and the result is 4. Thus the total of the balances ($600) divided by the number of customers (4) results in an average balance of $150!

NOTE

You can use an AS clause with a function. For example, the following command computes a sum of the BALANCE column and displays the column heading as TOTAL_BALANCE in the query results:

```
SELECT SUM(BALANCE) AS TOTAL_BALANCE
FROM CUSTOMER;
```

Using the DISTINCT Operator

In some situations, the **DISTINCT** operator is useful when used in conjunction with the COUNT function because it eliminates duplicate values in the query results. Examples 19 and 20 illustrate the most common uses of the DISTINCT operator.

EXAMPLE 19

Find the number of each customer that currently has an open order (that is, an order currently in the ORDERS table).

The command seems fairly simple. When a customer currently has an open order, there must be at least one row in the ORDERS table on which that customer's number appears. You could use the query shown in Figure 4-26 to find the customer numbers with open orders.

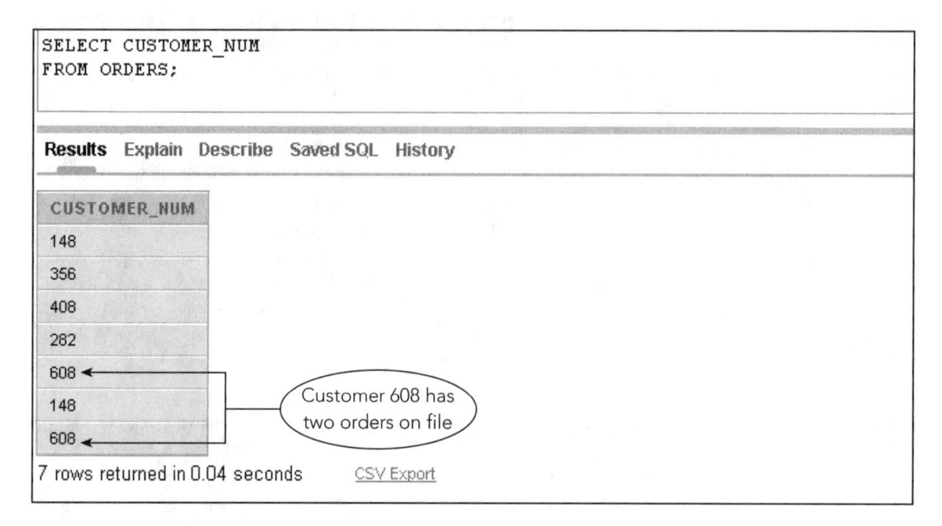

FIGURE 4-26 Numbers of customers with open orders

Notice that customer numbers 148 and 608 each appear more than once in the results; this means that both customers currently have more than one open order in the ORDERS table. Suppose you want to list each customer only once, as illustrated in Example 20.

EXAMPLE 20

Find the number of each customer that currently has an open order. List each customer only once.

To ensure uniqueness, you can use the DISTINCT operator, as shown in Figure 4-27.

```
SELECT DISTINCT(CUSTOMER_NUM)
FROM ORDERS;
```

Results Explain Describe Saved SQL History

CUSTOMER_NUM
282
148
608
356
408

5 rows returned in 0.04 seconds CSV Export

FIGURE 4-27 Numbers of customers with open orders and with duplicates removed

You might wonder about the relationship between COUNT and DISTINCT, because both involve counting rows. Example 21 identifies the differences.

EXAMPLE 21

Count the number of customers that currently have open orders.

The query shown in Figure 4-28 counts the number of customers using the CUSTOMER_NUM column.

```
SELECT COUNT(CUSTOMER_NUM)
FROM ORDERS;
```

Results Explain Describe Saved SQL History

COUNT(CUSTOMER_NUM)
7

1 rows returned in 0.01 seconds CSV Export

FIGURE 4-28 Count that includes duplicate customer numbers

Q & A

Question: What is wrong with the query results shown in Figure 4-28?
Answer: The answer, 7, is the result of counting the customers that have open orders multiple times—once for each separate order currently on file. The result counts each customer number and does not eliminate duplicate customer numbers to provide an accurate count of the number of customers.

Some SQL implementations, including Oracle and SQL Server (but not Access), allow you to use the DISTINCT operator to calculate the correct count, as shown in Figure 4-29.

```
SELECT COUNT(DISTINCT(CUSTOMER_NUM))
FROM ORDERS;
```

Results Explain Describe Saved SQL History

COUNT(DISTINCT(CUSTOMER_NUM))
5

1 rows returned in 0.04 seconds CSV Export

FIGURE 4-29 Count that excludes duplicate customer numbers (using DISTINCT within COUNT)

NESTING QUERIES

Sometimes obtaining the results you need requires two or more steps, as shown in the next two examples.

EXAMPLE 22

List the number of each part in class AP.

The command to obtain the answer is shown in Figure 4-30.

```
SELECT PART_NUM
FROM PART
WHERE CLASS = 'AP';
```

Results Explain Describe Saved SQL History

PART_NUM
CD52
DR93
DW11
KL62
KT03

5 rows returned in 0.02 seconds CSV Export

FIGURE 4-30 Selecting all parts in class AP

EXAMPLE 23

List the order numbers that contain an order line for a part in class AP.

Example 23 asks you to find the order numbers in the ORDER_LINE table that correspond to the part numbers in the results of the query used in Example 22. After viewing those results (CD52, DR93, DW11, KL62, and KT03), you can use the command shown in Figure 4-31.

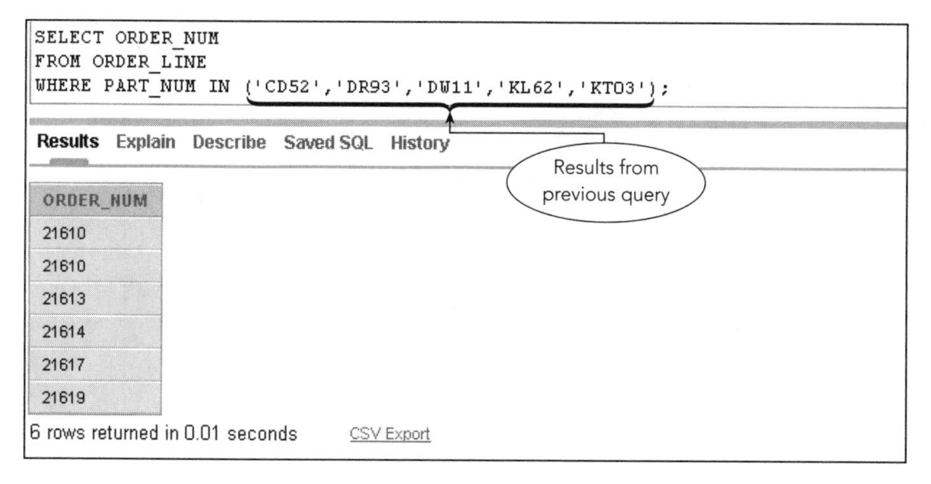

FIGURE 4-31 Query using the results from Figure 4-30

Subqueries

It is possible to place one query inside another. The inner query is called a **subquery**. The subquery is evaluated first. After the subquery has been evaluated, the outer query can use the results of the subquery to find its results, as shown in Example 24.

EXAMPLE 24

Find the answer to Examples 22 and 23 in one step.

You can find the same result as in the previous two examples in a single step by using a subquery. In Figure 4-32, the command shown in parentheses is the subquery. This subquery is evaluated first, producing a temporary table. The temporary table is used only to evaluate the query—it is not available to the user or displayed—and it is deleted after the evaluation of the query is complete. In this example, the temporary table has only a single column (PART_NUM) and five rows (CD52, DR93, DW11, KL62, and KT03). The outer query is evaluated next. In this case, the outer query retrieves the order number on every row in the ORDER_LINE table for which the part number is in the results of the subquery. Because that table contains only the part numbers in class AP, the results display the desired list of order numbers.

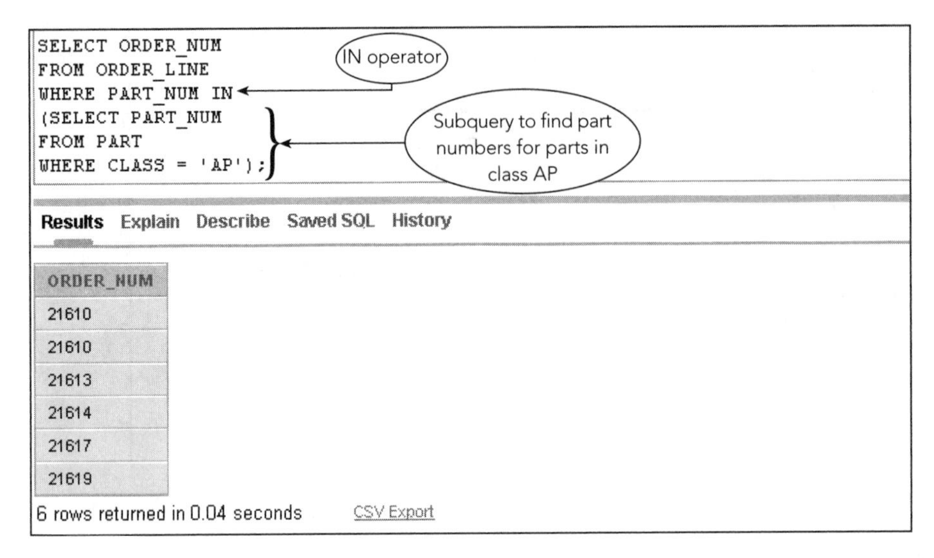

FIGURE 4-32 Using the IN operator and a subquery

Figure 4-32 shows duplicate order numbers in the results. To eliminate this duplication, you can use the DISTINCT operator as follows:

```
SELECT DISTINCT(ORDER_NUM)
FROM ORDER_LINE
WHERE PART_NUM IN
(SELECT PART_NUM
FROM PART
WHERE CLASS = 'AP');
```

The results of this query will display each order number only once.

EXAMPLE 25

List the number, name, and balance for each customer whose balance exceeds the average balance of all customers.

In this case, you use a subquery to obtain the average balance. Because this subquery produces a single number, you can compare each customer's balance with this number, as shown in Figure 4-33.

```
SELECT CUSTOMER_NUM, CUSTOMER_NAME, BALANCE
FROM CUSTOMER
WHERE BALANCE >
(SELECT AVG(BALANCE)
FROM CUSTOMER);
```

Results Explain Describe Saved SQL History

CUSTOMER_NUM	CUSTOMER_NAME	BALANCE
148	Al's Appliance and Sport	6550
356	Ferguson's	5785
408	The Everything Shop	5285.25
524	Kline's	12762
842	All Season	8221

5 rows returned in 0.01 seconds CSV Export

FIGURE 4-33 Query using an operator and a subquery

NOTE

You cannot use the condition BALANCE > AVG(BALANCE) in the WHERE clause; you must use a subquery to obtain the average balance. Then you can use the results of the subquery in a condition, as illustrated in Figure 4-33.

GROUPING

Grouping creates groups of rows that share some common characteristic. If you group customers by credit limit, for example, the first group contains customers with $5,000 credit limits, the second group contains customers with $7,500 credit limits, and so on. If, on the other hand, you group customers by sales rep number, the first group contains those customers represented by sales rep number 20, the second group contains those customers represented by sales rep number 35, and the third group contains those customers represented by sales rep number 65.

When you group rows, any calculations indicated in the SELECT command are performed for the entire group. For example, if you group customers by rep number and the query requests the average balance, the results include the average balance for the group of customers represented by rep number 20, the average balance for the group represented by rep number 35, and the average balance for the group represented by rep number 65. The following examples illustrate this process.

Using the GROUP BY Clause

The **GROUP BY clause** lets you group data on a particular column, such as REP_NUM, and then calculate statistics, when desired, as shown in Example 26.

EXAMPLE 26

For each sales rep, list the rep number and the average balance of the rep's customers.

Because you need to group customers by rep number and then calculate the average balance for all customers in each group, you must use the GROUP BY clause. In this case, GROUP BY REP_NUM puts customers with the same rep number into separate groups. Any statistics indicated in the SELECT command are calculated for each group. It is important to note that the GROUP BY clause does not sort the data in a particular order; you must use the ORDER BY clause to sort data. Assuming that the results should be ordered by rep number, you can use the command shown in Figure 4-34.

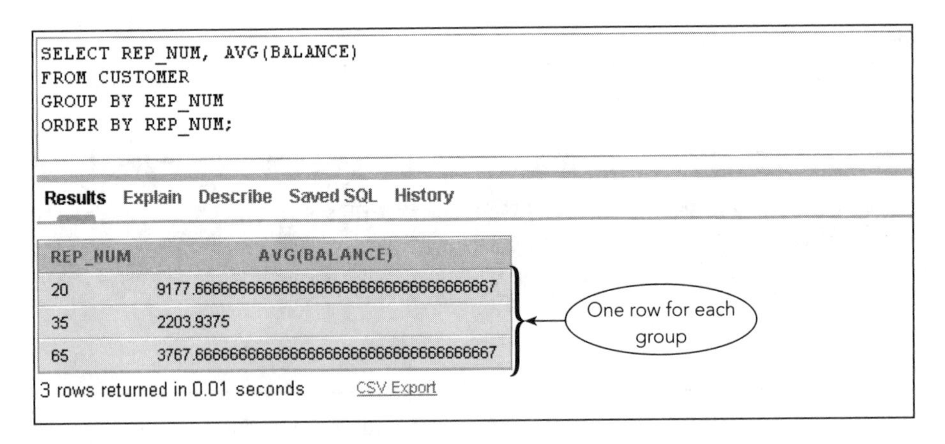

FIGURE 4-34 Grouping records on a column

When rows are grouped, one line of output is produced for each group. The only things that can be displayed are statistics calculated for the group or columns whose values are the same for all rows in a group.

Q & A

Question: Is it appropriate to display the rep number in the query for Example 26?
Answer: Yes, because the rep number in one row in a group must be the same as the rep number in any other row in the group.

Q & A

Question: Would it be appropriate to display a customer number in the query for Example 26?
Answer: No, because the customer number varies on the rows in a group. (The same rep is associated with many customers.) The DBMS would not be able to determine which customer number to display for the group, and would display an error message if you attempt to display a customer number.

Using a HAVING Clause

The HAVING clause is used to restrict the groups that are included, as shown in Example 27.

EXAMPLE 27

Repeat the previous example, but list only those reps who represent fewer than four customers.

The only difference between Examples 26 and 27 is the restriction to display only those reps who represent fewer than four customers. This restriction does not apply to individual rows but rather to *groups*. Because the WHERE clause applies only to rows, you cannot use it to accomplish the kind of selection that is required. Fortunately, the HAVING clause does for groups what the WHERE clause does for rows. The **HAVING clause** limits the groups that are included in the results. In Figure 4-35, the row created for a group is displayed only when the count of the number of rows in the group is less than four; in addition, all groups are ordered by rep number.

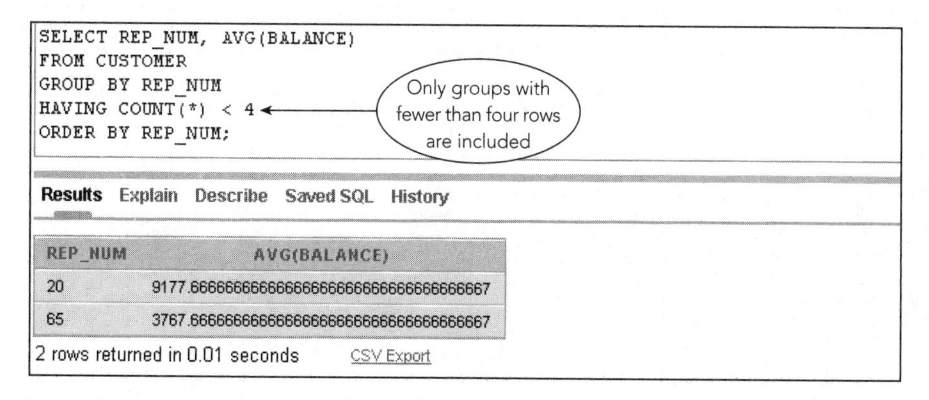

FIGURE 4-35 Restricting the groups to include in the results

HAVING vs. WHERE

Just as you can use the WHERE clause to limit the *rows* that are included in a query's result, you can use the HAVING clause to limit the *groups* that are included. The following examples illustrate the difference between these two clauses.

EXAMPLE 28

List each credit limit and the number of customers having each credit limit.

To count the number of customers that have a given credit limit, you must group the data by credit limit, as shown in Figure 4-36.

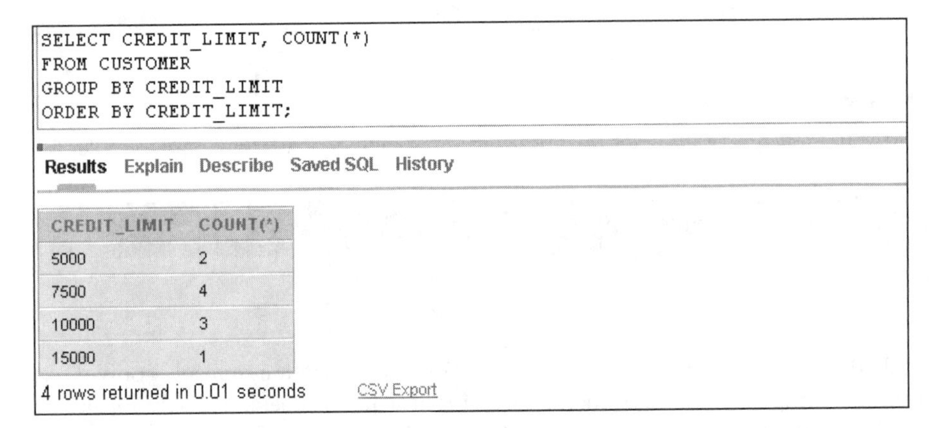

```
SELECT CREDIT_LIMIT, COUNT(*)
FROM CUSTOMER
GROUP BY CREDIT_LIMIT
ORDER BY CREDIT_LIMIT;
```

Results Explain Describe Saved SQL History

CREDIT_LIMIT	COUNT(*)
5000	2
7500	4
10000	3
15000	1

4 rows returned in 0.01 seconds CSV Export

FIGURE 4-36 Counting the number of rows in each group

EXAMPLE 29

Repeat Example 28, but list only those credit limits held by more than one customer.

Because this condition involves a group total, the query includes a HAVING clause, as shown in Figure 4-37.

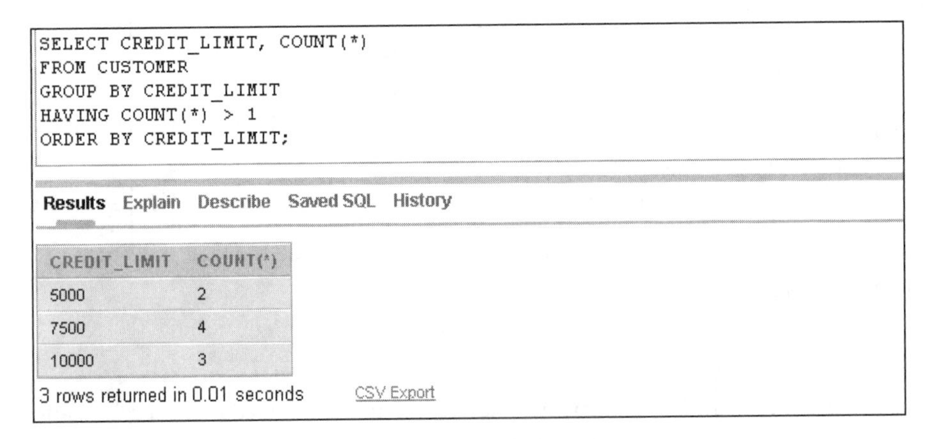

```
SELECT CREDIT_LIMIT, COUNT(*)
FROM CUSTOMER
GROUP BY CREDIT_LIMIT
HAVING COUNT(*) > 1
ORDER BY CREDIT_LIMIT;
```

Results Explain Describe Saved SQL History

CREDIT_LIMIT	COUNT(*)
5000	2
7500	4
10000	3

3 rows returned in 0.01 seconds CSV Export

FIGURE 4-37 Displaying groups that contain more than one row

EXAMPLE 30

List each credit limit and the number of customers of sales rep 20 that have this limit.

The condition involves only rows, so using the WHERE clause is appropriate, as shown in Figure 4-38.

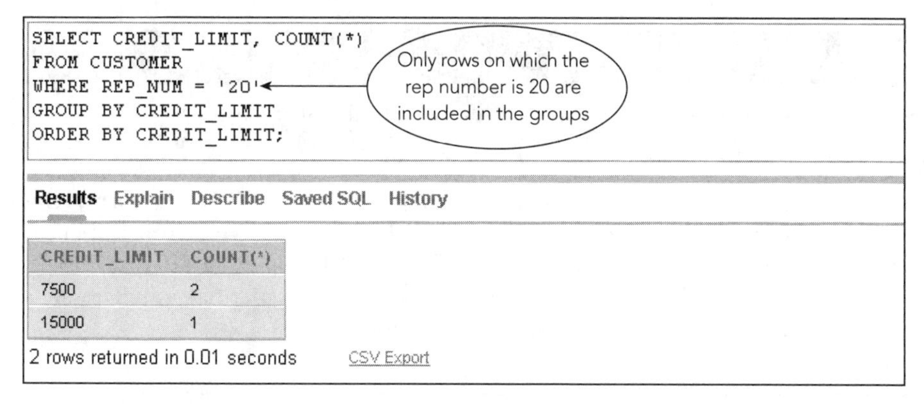

```
SELECT CREDIT_LIMIT, COUNT(*)
FROM CUSTOMER
WHERE REP_NUM = '20'
GROUP BY CREDIT_LIMIT
ORDER BY CREDIT_LIMIT;
```


CREDIT_LIMIT	COUNT(*)
7500	2
15000	1

2 rows returned in 0.01 seconds CSV Export

FIGURE 4-38 Restricting the rows to be grouped

EXAMPLE 31

Repeat Example 30, but list only those credit limits held by more than one customer.

Because the conditions involve rows and groups, you must use both a WHERE clause and a HAVING clause, as shown in Figure 4-39.

```
SELECT CREDIT_LIMIT, COUNT(*)
FROM CUSTOMER
WHERE REP_NUM = '20'
GROUP BY CREDIT_LIMIT
HAVING COUNT(*) > 1
ORDER BY CREDIT_LIMIT;
```

Results Explain Describe Saved SQL History

CREDIT_LIMIT	COUNT(*)
7500	2

1 rows returned in 0.01 seconds CSV Export

FIGURE 4-39 Restricting the rows and the groups

In Example 31, rows from the original table are evaluated only when the sales rep number is 20. These rows then are grouped by credit limit and the count is calculated. Only groups for which the calculated count is greater than one are displayed.

NULLS

Sometimes a condition involves a column that can accept null values, as illustrated in Example 32.

EXAMPLE 32

List the number and name of each customer with a null (unknown) street value.

You might expect the condition to be something like STREET = NULL. The correct format actually uses the **IS NULL** operator (STREET IS NULL), as shown in Figure 4-40. (To select a customer whose street is not null, use the **IS NOT NULL** operator (STREET IS NOT NULL).) In the current Premiere Products database, no customer has a null street value; therefore, no rows are retrieved in the query results.

```
SELECT CUSTOMER_NUM, CUSTOMER_NAME
FROM CUSTOMER
WHERE STREET IS NULL;
```

Results Explain Describe Saved SQL History

no data found

FIGURE 4-40 Selecting rows containing null values in the STREET column

SUMMARY OF SQL CLAUSES, FUNCTIONS, AND OPERATORS

In this chapter, you learned how to create queries that retrieve data from a single table by constructing appropriate SELECT commands. In the next chapter, you will learn how to create queries that retrieve data from multiple tables. The queries you created in this chapter used the clauses, functions, and operators shown in Figure 4-41.

Clause, function, or operator	Description
AND operator	Specifies that all simple conditions must be true for the compound condition to be true
AVG function	Calculates the average value in a numeric range
BETWEEN operator	Specifies a range of values in a condition
COUNT function	Counts the number of rows in a table
DESC operator	Sorts the query results in descending order based on the column name
DISTINCT operator	Ensures uniqueness in the condition by eliminating redundant values
FROM clause	Indicates the table from which to retrieve the specified columns
GROUP BY clause	Groups rows based on the specified column
HAVING clause	Limits a condition to the groups that are included
IN clause	Uses the IN operator to find a value in a group of values specified in the condition
IS NOT NULL operator	Finds rows that do not contain a null value in the specified column
IS NULL operator	Finds rows that contain a null value in the specified column
LIKE operator	Indicates a pattern of characters to find in a condition
MAX function	Calculates the maximum value in a numeric range
MIN function	Calculates the minimum value in a numeric range
NOT operator	Reverses the truth or falsity of the original condition
OR operator	Specifies that the compound condition is true whenever any of the simple conditions is true
ORDER BY clause	Lists the query results in the specified order based on the column name
SELECT clause	Specifies the columns to retrieve in the query
SUM function	Totals the values in a numeric range
WHERE clause	Specifies any conditions for the query

FIGURE 4-41 SQL query clauses, functions, and operators

Chapter Summary

- The basic form of the SQL SELECT command is SELECT-FROM-WHERE. Specify the columns to be listed after the word SELECT (or type an asterisk (*) to select all columns), and then specify the table name that contains these columns after the word FROM. Optionally, you can include one or more conditions after the word WHERE.
- Simple conditions are written in the following form: column name, comparison operator, column name or value. Simple conditions can involve any of the comparison operators: =, >, >=, <, <=, < >, or !=.
- You can form compound conditions by combining simple conditions using the AND, OR, and NOT operators.
- Use the BETWEEN operator to indicate a range of values in a condition.
- Use computed columns in SQL commands by using arithmetic operators and writing the computation in place of a column name. You can assign a name to the computed column by following the computation with the word AS and then the desired name.
- To check for a value in a character column that is similar to a particular string of characters, use the LIKE operator. In Oracle and SQL Server, the percent (%) wildcard represents any collection of characters, and the underscore (_) wildcard represents any single character. In Access, the asterisk (*) wildcard represents any collection of characters, and the question mark (?) wildcard represents any single character.
- To determine whether a column contains a value in a set of values, use the IN operator.
- Use an ORDER BY clause to sort data. List sort keys in order of importance. To sort in descending order, follow the sort key with the DESC operator.
- SQL processes the aggregate functions COUNT, SUM, AVG, MAX, and MIN. These calculations apply to groups of rows.
- To avoid duplicates in a query that uses an aggregate function, precede the column name with the DISTINCT operator.
- When one SQL query is placed inside another, it is called a subquery. The inner query (the subquery) is evaluated first.
- Use a GROUP BY clause to group data.
- Use a HAVING clause to restrict the output to certain groups.
- Use the IS NULL operator in a WHERE clause to find rows containing a null value in a particular column. Use the IS NOT NULL operator in a WHERE clause to find rows that do not contain a null value.

Key Terms

aggregate function
AND
AND condition
AVG
BETWEEN
compound condition
computed column
COUNT
DESC
DISTINCT

FROM clause
GROUP BY clause
grouping
HAVING clause
IN clause
IS NOT NULL
IS NULL
key
LIKE
major sort key
MAX
MIN
minor sort key
NOT
NOT condition
OR
OR condition
ORDER BY clause
primary sort key
query
secondary sort key
SELECT clause
simple condition
sort key
subquery
SUM
WHERE clause

Review Questions

1. Describe the basic form of the SQL SELECT command.
2. How do you form a simple condition?
3. How do you form a compound condition?
4. In SQL, what operator do you use to determine whether a value is between two other values without using an AND condition?
5. How do you use a computed column in SQL? How do you name the computed column?
6. In which clause would you use a wildcard in a condition?
7. What wildcards are available in Oracle, and what do they represent?
8. How do you determine whether a column contains one of a particular set of values without using an AND condition?
9. How do you sort data?
10. How do you sort data on more than one sort key? What is the more important key called? What is the less important key called?
11. How do you sort data in descending order?
12. What are the SQL aggregate functions?
13. How do you avoid including duplicate values in a query's results?
14. What is a subquery?
15. How do you group data in an SQL query?
16. When grouping data in a query, how do you restrict the output to only those groups satisfying some condition?
17. How do you find rows in which a particular column contains a null value?

18. Use your favorite Web browser and Web search engine to find out how to enter a date in an SQL query in Oracle, Access, and SQL Server. Using the information you find, complete the following SQL command for each of the three DBMSs (Oracle, Access, and SQL Server) to list orders placed on October 20, 2010:

```
SELECT *
FROM ORDERS
WHERE ORDER_DATE =
```

Be sure to reference the URLs that contain the information.

Exercises

Premiere Products

Use SQL and the Premiere Products database (see Figure 1-2 in Chapter 1) to complete the following exercises. If directed to do so by your instructor, use the information provided with the Chapter 3 Exercises to print your output.

1. List the part number, description, and price for all parts.
2. List all rows and columns for the complete ORDERS table.
3. List the names of customers with credit limits of $10,000 or more.
4. List the order number for each order placed by customer number 608 on 10/23/2010. (*Hint:* If you need help, use the discussion of the DATE data type in Figure 3-11 in Chapter 3.)
5. List the number and name of each customer represented by sales rep 35 or sales rep 65.
6. List the part number and part description of each part that is not in item class AP.
7. List the part number, description, and number of units on hand for each part that has between 10 and 25 units on hand, including both 10 and 25. Do this two ways.
8. List the part number, part description, and on-hand value (units on hand * unit price) of each part in item class SG. (On-hand value is really units on hand * cost, but there is no COST column in the PART table.) Assign the name ON_HAND_VALUE to the computed column.
9. List the part number, part description, and on-hand value for each part whose on-hand value is at least $7,500. Assign the name ON_HAND_VALUE to the computed column.
10. Use the IN operator to list the part number and part description of each part in item class AP or SG.
11. Find the number and name of each customer whose name begins with the letter "B."
12. List all details about all parts. Order the output by part description.
13. List all details about all parts. Order the output by part number within warehouse. (That is, order the output by warehouse and then by part number.)
14. How many customers have balances that are more than their credit limits?
15. Find the total of the balances for all customers represented by sales rep 65 with balances that are less than their credit limits.
16. List the part number, part description, and on-hand value of each part whose number of units on hand is more than the average number of units on hand for all parts. (*Hint:* Use a subquery.)

17. What is the price of the least expensive part in the database?
18. What is the part number, description, and price of the least expensive part in the database? (*Hint:* Use a subquery.)
19. List the sum of the balances of all customers for each sales rep. Order and group the results by sales rep number.
20. List the sum of the balances of all customers for each sales rep, but restrict the output to those sales reps for which the sum is more than $10,000.
21. List the part number of any part with an unknown description.

Henry Books

Use SQL and the Henry Books database (Figures 1-4 through 1-7 in Chapter 1) to complete the following exercises. If directed to do so by your instructor, use the information provided with the Chapter 3 Exercises to print your output.

1. List the book code and book title of each book.
2. List the complete BRANCH table.
3. List the name of each publisher located in Boston.
4. List the name of each publisher not located in New York.
5. List the name of each branch that has at least nine employees.
6. List the book code and book title of each book that has the type HOR.
7. List the book code and book title of each book that has the type HOR and is in paperback.
8. List the book code and book title of each book that has the type HOR or is published by the publisher with the publisher code SC.
9. List the book code, book title, and price of each book with a price between $15 and $25.
10. List the book code and book title of each book that has the type MYS and a price of less than $20.
11. Customers who are part of a special program get a 10 percent discount off regular book prices. List the book code, book title, and discounted price of each book. Use DISCOUNTED_PRICE as the name for the computed column, which should calculate 90 percent of the current price (100 percent less a 10 percent discount).
12. Find the name of each publisher containing the word "and." (*Hint:* Be sure that your query selects only those publishers that contain the word "and" and not those that contain the letters "and" in the middle of a word. For example, your query should select the publisher named "Farrar Straus and Giroux," but should *not* select the publisher named "Random House.")
13. List the book code and book title of each book that has the type SFI, MYS, or HOR. Use the IN operator in your command.
14. Repeat Exercise 13, but also list the books in alphabetical order by title.
15. Repeat Exercise 13, but also include the price, and list the books in descending order by price. Within a group of books having the same price, further order the books by title.
16. Display the list of book types in the database. List each book type only once.
17. How many books have the type SFI?

18. For each type of book, list the type and the average price.
19. Repeat Exercise 18, but consider only paperback books.
20. Repeat Exercise 18, but consider only paperback books for those types for which the average price is more than $10.
21. What are the title(s) and price(s) of the least expensive book(s) in the database?
22. What is the most expensive book in the database?
23. How many employees does Henry Books have?

Alexamara Marina Group

Use SQL and the Alexamara Marina Group database (Figures 1-8 through 1-12 in Chapter 1) to complete the following exercises. If directed to do so by your instructor, use the information provided with the Chapter 3 Exercises to print your output.

1. List the owner number, last name, and first name of every boat owner.
2. List the complete MARINA table (all rows and all columns).
3. List the last name and first name of every owner who lives in Rivard.
4. List the last name and first name of every owner who does not live in Rivard.
5. List the marina number and slip number for every slip whose length is equal to or less than 30 feet.
6. List the marina number and slip number for every boat with the type Ray 4025.
7. List the slip number for every boat with the type Ray 4025 that is located in marina 1.
8. List the boat name for each boat located in a slip whose length is between 25 and 30 feet.
9. List the slip number for every slip in marina 1 whose rental fee is less than $3,000.
10. Labor is billed at the rate of $60 per hour. List the slip ID, category number, estimated hours, and estimated labor cost for every service request. To obtain the estimated labor cost, multiply the estimated hours by 60. Use the column name ESTIMATED_COST for the estimated labor cost.
11. List the marina number and slip number for all slips containing a boat with the type Sprite 4000, Sprite 3000, or Ray 4025.
12. List the marina number, slip number, and boat name for all boats. Sort the results by boat name within marina number.
13. How many Dolphin 28 boats are stored at both marinas?
14. Calculate the total rental fees Alexamara receives each year based on the length of the slip.

CHAPTER 5

MULTIPLE-TABLE QUERIES

LEARNING OBJECTIVES

Objectives

- Use joins to retrieve data from more than one table
- Use the IN and EXISTS operators to query multiple tables
- Use a subquery within a subquery
- Use an alias
- Join a table to itself
- Perform set operations (union, intersection, and difference)
- Use the ALL and ANY operators in a query
- Perform special operations (inner join, outer join, and product)

INTRODUCTION

In this chapter, you will learn how to use SQL to retrieve data from two or more tables using one SQL command. You will join tables together and examine how to obtain similar results using the SQL IN and EXISTS operators. Then you will use aliases to simplify queries and join a table to itself. You also will implement the set operations of union, intersection, and difference using SQL commands. You will examine two related SQL operators: ALL and ANY. Finally, you will perform inner joins, outer joins, and products.

QUERYING MULTIPLE TABLES

In Chapter 4, you learned how to retrieve data from a single table. Many queries require you to retrieve data from two or more tables. To retrieve data from multiple tables, you first must join the tables, and then formulate a query using the same commands that you use for single-table queries.

NOTE

In the following queries, your results might contain the same rows, but they might be listed in a different order. If order is important, you can include an ORDER BY clause in the query to ensure that the results are listed in the desired order.

Joining Two Tables

To retrieve data from more than one table, you must **join** the tables together by finding rows in the two tables that have identical values in matching columns. You can join tables by using a condition in the WHERE clause, as you will see in Example 1.

EXAMPLE 1

List the number and name of each customer, together with the number, last name, and first name of the sales rep who represents the customer.

Because the customer numbers and names are in the CUSTOMER table and the sales rep numbers and names are in the REP table, you need to include both tables in the SQL command so you can retrieve data from both tables. To join (relate) the tables, you construct the SQL command as follows:

1. In the SELECT clause, list all columns you want to display.
2. In the FROM clause, list all tables involved in the query.
3. In the WHERE clause, list the condition that restricts the data to be retrieved to only those rows from the two tables that match; that is, restrict it to the rows that have common values in matching columns.

As you learned in Chapter 2, it is often necessary to qualify a column name to specify the particular column you are referencing. Qualifying column names is especially important when joining tables because you must join tables on *matching* columns that frequently have identical column names. To qualify a column name, precede the name of the column with the name of the table, followed by a period. The matching columns in this example are both named REP_NUM—there is a column in the REP table named REP_NUM and a column in the CUSTOMER table that also is named REP_NUM. The REP_NUM column in the REP table is written as REP.REP_NUM and the REP_NUM column in the CUSTOMER table is written as CUSTOMER.REP_NUM. The query and its results appear in Figure 5-1.

```
SELECT CUSTOMER_NUM, CUSTOMER_NAME, REP.REP_NUM, LAST_NAME, FIRST_NAME
FROM CUSTOMER, REP
WHERE CUSTOMER.REP_NUM = REP.REP_NUM;
```

Condition to relate the tables

Results Explain Describe Saved SQL History

CUSTOMER_NUM	CUSTOMER_NAME	REP_NUM	LAST_NAME	FIRST_NAME
148	Al's Appliance and Sport	20	Kaiser	Valerie
282	Brookings Direct	35	Hull	Richard
356	Ferguson's	65	Perez	Juan
408	The Everything Shop	35	Hull	Richard
462	Bargains Galore	65	Perez	Juan
524	Kline's	20	Kaiser	Valerie
608	Johnson's Department Store	65	Perez	Juan
687	Lee's Sport and Appliance	35	Hull	Richard
725	Deerfield's Four Seasons	35	Hull	Richard
842	All Season	20	Kaiser	Valerie

10 rows returned in 0.70 seconds CSV Export

FIGURE 5-1 Joining two tables with a single SQL command

When there is potential ambiguity in listing column names, you *must* qualify the columns involved in the query. It is permissible to qualify other columns as well, even when there is no possible confusion. Some people prefer to qualify all column names; in this text, however, you will qualify column names only when necessary.

Q & A

Question: In the first row of output in Figure 5-1, the customer number is 148, and the customer name is Al's Appliance and Sport. These values represent the first row of the CUSTOMER table. Why is the sales rep number 20, the last name of the sales rep Kaiser, and the first name Valerie?
Answer: In the CUSTOMER table, the sales rep number for customer number 148 is 20. (This indicates that customer number 148 is *related* to sales rep number 20.) In the REP table, the last name of sales rep number 20 is Kaiser and the first name is Valerie.

EXAMPLE 2

List the number and name of each customer whose credit limit is $7,500, together with the number, last name, and first name of the sales rep who represents the customer.

In Example 1, you used a condition in the WHERE clause only to relate a customer with a sales rep to join the tables. Although relating a customer with a sales rep is essential in this example as well, you also need to restrict the output to only those customers whose credit limits are $7,500. You can restrict the rows by using a compound condition, as shown in Figure 5-2.

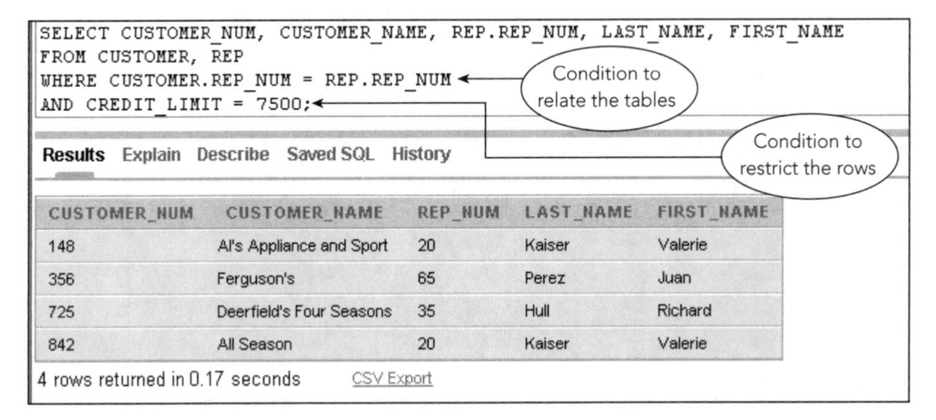

```
SELECT CUSTOMER_NUM, CUSTOMER_NAME, REP.REP_NUM, LAST_NAME, FIRST_NAME
FROM CUSTOMER, REP
WHERE CUSTOMER.REP_NUM = REP.REP_NUM
AND CREDIT_LIMIT = 7500;
```


CUSTOMER_NUM	CUSTOMER_NAME	REP_NUM	LAST_NAME	FIRST_NAME
148	Al's Appliance and Sport	20	Kaiser	Valerie
356	Ferguson's	65	Perez	Juan
725	Deerfield's Four Seasons	35	Hull	Richard
842	All Season	20	Kaiser	Valerie

FIGURE 5-2 Restricting the rows in a join

EXAMPLE 3

For every part on order, list the order number, part number, part description, number of units ordered, quoted price, and unit price.

A part is considered "on order" when there is a row in the ORDER_LINE table in which the part appears. You can find the order number, number of units ordered, and quoted price in the ORDER_LINE table. To find the part description and the unit price, however, you need to look in the PART table. Then you need to find rows in the ORDER_LINE table and rows in the PART table that match (rows containing the same part number). The query and its results appear in Figure 5-3.

```
SELECT ORDER_NUM, ORDER_LINE.PART_NUM, DESCRIPTION, NUM_ORDERED, QUOTED_PRICE, PRICE
FROM ORDER_LINE, PART
WHERE ORDER_LINE.PART_NUM = PART.PART_NUM;
```

Results Explain Describe Saved SQL History

ORDER_NUM	PART_NUM	DESCRIPTION	NUM_ORDERED	QUOTED_PRICE	PRICE
21608	AT94	Iron	11	21.95	24.95
21617	BV06	Home Gym	2	794.95	794.95
21617	CD52	Microwave Oven	4	150	165
21619	DR93	Gas Range	1	495	495
21610	DR93	Gas Range	1	495	495
21610	DW11	Washer	1	399.99	399.99
21613	KL62	Dryer	4	329.95	349.95
21614	KT03	Dishwasher	2	595	595
21623	KV29	Treadmill	2	1290	1390

9 rows returned in 0.06 seconds CSV Export

FIGURE 5-3 Joining the ORDER_LINE and PART tables

Q & A

Question: Can you use PART.PART_NUM instead of ORDER_LINE.PART_NUM in the SELECT clause?
Answer: Yes. The values for these two columns match because they must satisfy the condition ORDER_LINE.PART_NUM = PART.PART_NUM.

COMPARING JOINS, IN, AND EXISTS

You join tables in SQL by including a condition in the WHERE clause to ensure that matching columns contain equal values (for example, ORDER_LINE.PART_NUM = PART.PART_NUM). You can obtain similar results by using either the IN operator (described in Chapter 4) or the **EXISTS** operator with a subquery. The choice is a matter of personal preference because either approach obtains the same results. The following examples illustrate the use of each operator.

EXAMPLE 4

Find the description of each part included in order number 21610.

Because this query also involves retrieving data from the ORDER_LINE and PART tables (as illustrated in Example 3), you could approach it in a similar fashion. There are two basic differences, however, between Examples 3 and 4. First, the query in Example 4 does not require as many columns; second, it involves only order number 21610. Having fewer columns to retrieve means that there will be fewer columns listed in the SELECT clause. You can restrict the query to a single order by adding the condition ORDER_NUM = '21610' to the WHERE clause. The query and its results appear in Figure 5-4.

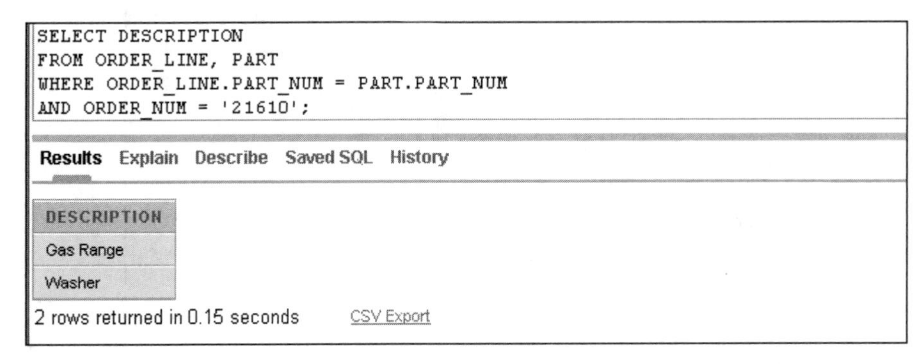

FIGURE 5-4 Restricting the rows when joining the ORDER_LINE and PART tables

Notice that the ORDER_LINE table is listed in the FROM clause, even though you do not need to display any columns from the ORDER_LINE table. The WHERE clause contains columns from the ORDER_LINE table, so it is necessary to include the table in the FROM clause.

Using the IN Operator

Another way to retrieve data from multiple tables in a query is to use the IN operator with a subquery. In Example 4, you first could use a subquery to find all part numbers in the ORDER_LINE table that appear in any row on which the order number is 21610. Then you could find the part description for any part whose part number is in this list. The query and its results appear in Figure 5-5.

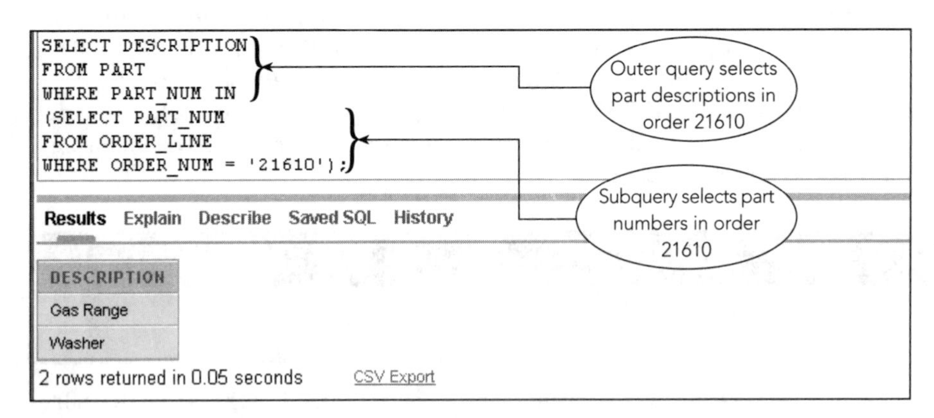

FIGURE 5-5 Using the IN operator instead of a join to query two tables

In Figure 5-5, evaluating the subquery produces a temporary table consisting of those part numbers (DR93 and DW11) that are present in order number 21610. Executing the remaining portion of the query produces part descriptions for each part whose number is in this temporary table; in this case, Gas Range (DR93) and Washer (DW11).

Using the EXISTS Operator

You also can use the EXISTS operator to retrieve data from more than one table, as shown in Example 5. The **EXISTS** operator checks for the existence of rows that satisfy some criterion.

EXAMPLE 5

Find the order number and order date for each order that contains part number DR93.

This query is similar to the one in Example 4, but this time the query involves the ORDERS table and not the PART table. In this case, you can write the query in either of the ways previously demonstrated. For example, you could use the IN operator with a subquery, as shown in Figure 5-6.

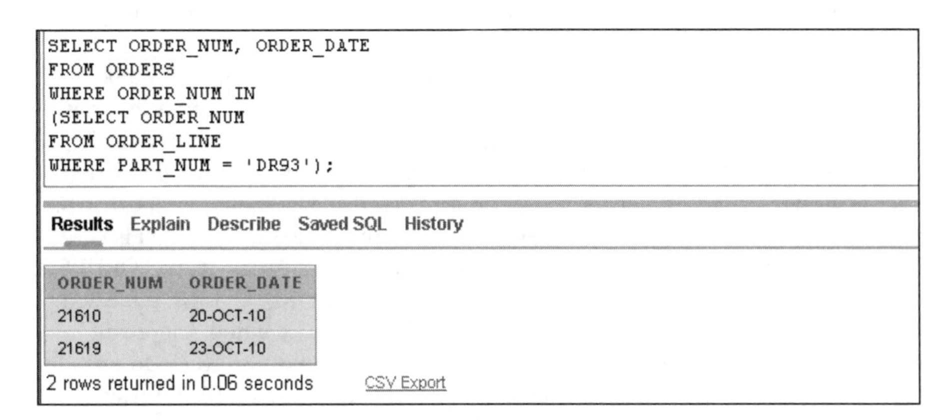

FIGURE 5-6 Using the IN operator to select order information

Using the EXISTS operator provides another approach to solving Example 5, as shown in Figure 5-7.

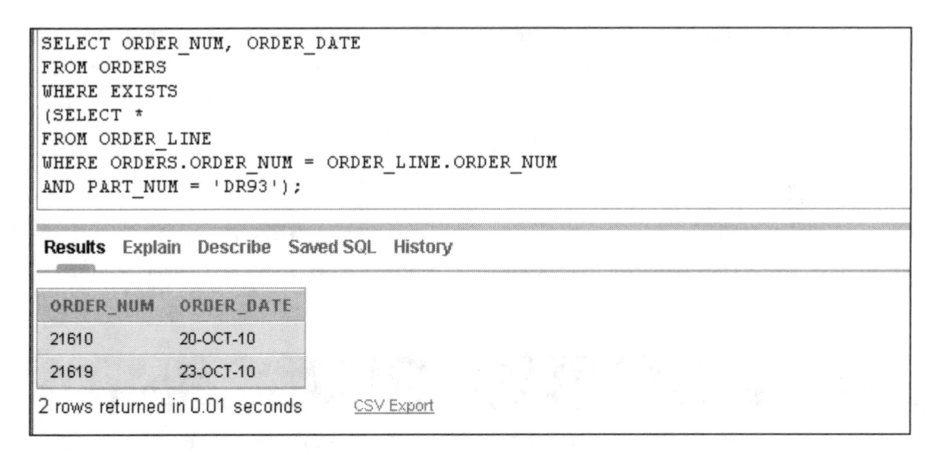

FIGURE 5-7 Using the EXISTS operator to select order information

The subquery in Figure 5-7 is the first one you have seen that involves a table listed in the outer query. This type of subquery is called a **correlated subquery**. In this case, the ORDERS table, which is listed in the FROM clause of the outer query, is used in the subquery. For this reason, you need to qualify the ORDER_NUM column in the subquery (ORDERS.ORDER_NUM). You did not need to qualify the columns in the previous queries involving the IN operator.

The query shown in Figure 5-7 works as follows. For each row in the ORDERS table, the subquery is executed using the value of ORDERS.ORDER_NUM that occurs in that row. The inner query produces a list of all rows in the ORDER_LINE table in which ORDER_LINE.ORDER_NUM matches this value and in which PART_NUM is equal to DR93. You can precede a subquery with the EXISTS operator to create a condition that is true if one or more rows are obtained when the subquery is executed; otherwise, the condition is false.

To illustrate the process, consider order numbers 21610 and 21613 in the ORDERS table. Order number 21610 is included because a row exists in the ORDER_LINE table with this order number and part number DR93. When the subquery is executed, there will be at least one row in the results, which in turn makes the EXISTS condition true. Order number 21613, however, will not be included because no row exists in the ORDER_LINE table with this order number and part number DR93. There will be no rows contained in the results of the subquery, which in turn makes the EXISTS condition false.

Using a Subquery Within a Subquery

You can use SQL to create a **nested subquery** (a subquery within a subquery), as illustrated in Example 6.

EXAMPLE 6

Find the order number and order date for each order that includes a part located in warehouse 3.

One way to approach this problem is first to determine the list of part numbers in the PART table for each part located in warehouse 3. Then you obtain a list of order numbers in the ORDER_LINE table with a corresponding part number in the part number list. Finally, you retrieve those order numbers and order dates in the ORDERS table for which the order number is in the list of order numbers obtained during the second step. The query and its results appear in Figure 5-8.

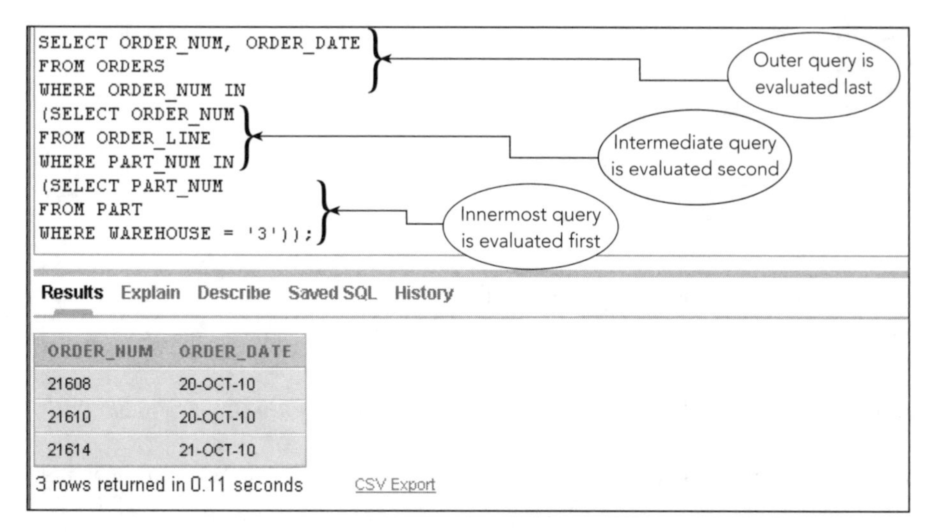

ORDER_NUM	ORDER_DATE
21608	20-OCT-10
21610	20-OCT-10
21614	21-OCT-10

FIGURE 5-8 Nested subqueries (a subquery within a subquery)

As you might expect, SQL evaluates the queries from the innermost query to the outermost query. The query in this example is evaluated in three steps:

1. The innermost subquery is evaluated first, producing a temporary table of part numbers for those parts located in warehouse 3.
2. The next (intermediate) subquery is evaluated, producing a second temporary table with a list of order numbers. Each order number in this collection has a row in the ORDER_LINE table for which the part number is in the temporary table produced in Step 1.
3. The outer query is evaluated last, producing the desired list of order numbers and order dates. Only those orders whose numbers are in the temporary table produced in Step 2 are included in the results.

Another approach to solving Example 6 involves joining the ORDERS, ORDER_LINE, and PART tables. The query and its results appear in Figure 5-9.

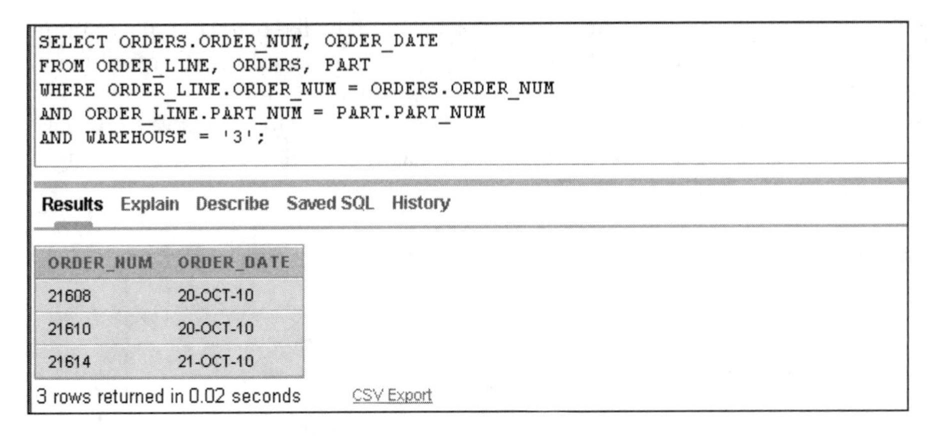

FIGURE 5-9 Joining three tables

In this query, the following conditions join the tables:

```
ORDER_LINE.ORDER_NUM = ORDERS.ORDER_NUM
ORDER_LINE.PART_NUM = PART.PART_NUM
```

The condition WAREHOUSE = '3' restricts the output to only those parts located in warehouse 3.

The query results are correct regardless of which command you use. You can use whichever approach you prefer.

You might wonder whether one approach is more efficient than the other. SQL performs many built-in optimizations that analyze queries to determine the best way to satisfy them. Given a good optimizer, it should not make much difference how you formulate the query—you can see that using nested subqueries (Figure 5-8) produces the query in 0.11 seconds and joining the tables (Figure 5-9) produces the results in 0.02 seconds. If you are using a DBMS without an optimizer, however, the way you write a query *can* make a difference in the speed at which the DBMS executes the query. When you are working with a very large database and efficiency is a prime concern, consult the DBMS's manual or try some timings yourself. Try running the same query both ways to see whether you notice a difference in the speed of execution. In small databases, there should not be a significant time difference between the two approaches.

A Comprehensive Example

The query used in Example 7 involves several of the features already presented. The query illustrates all the major clauses that you can use in a SELECT command. It also illustrates the order in which these clauses must appear.

EXAMPLE 7

List the customer number, order number, order date, and order total for each order with a total that exceeds $1,000. Assign the column name ORDER_TOTAL to the column that displays order totals.

The query and its results appear in Figure 5-10.

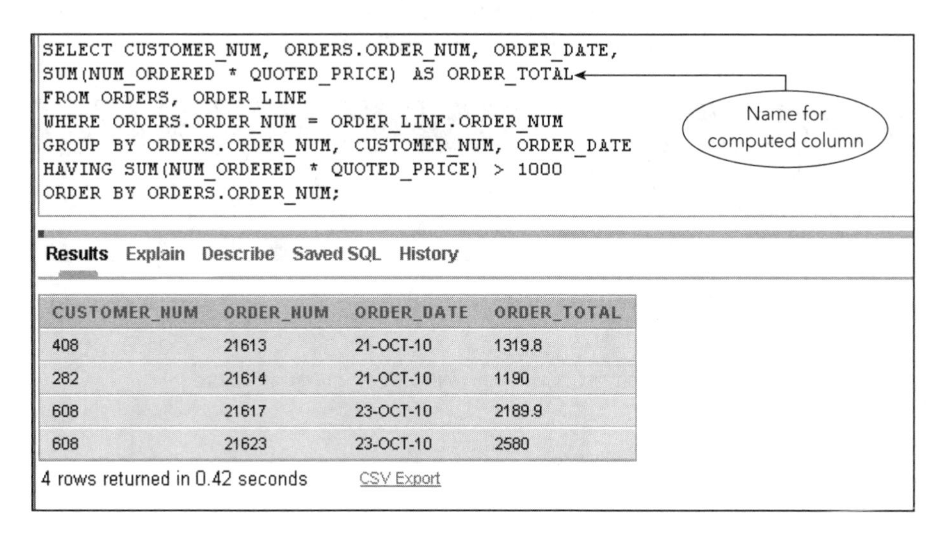

```
SELECT CUSTOMER_NUM, ORDERS.ORDER_NUM, ORDER_DATE,
SUM(NUM_ORDERED * QUOTED_PRICE) AS ORDER_TOTAL
FROM ORDERS, ORDER_LINE
WHERE ORDERS.ORDER_NUM = ORDER_LINE.ORDER_NUM
GROUP BY ORDERS.ORDER_NUM, CUSTOMER_NUM, ORDER_DATE
HAVING SUM(NUM_ORDERED * QUOTED_PRICE) > 1000
ORDER BY ORDERS.ORDER_NUM;
```


CUSTOMER_NUM	ORDER_NUM	ORDER_DATE	ORDER_TOTAL
408	21613	21-OCT-10	1319.8
282	21614	21-OCT-10	1190
608	21617	23-OCT-10	2189.9
608	21623	23-OCT-10	2580

4 rows returned in 0.42 seconds   CSV Export

FIGURE 5-10 Comprehensive example

In this query, the ORDERS and ORDER_LINE tables are joined by listing both tables in the FROM clause and relating them in the WHERE clause. Selected data is sorted by order number using the ORDER BY clause. The GROUP BY clause indicates that the data is to be grouped by order number, customer number, and order date. For each group, the SELECT clause displays the customer number, order number, order date, and order total (SUM(NUM_ORDERED * QUOTED_PRICE)). In addition, the total was renamed ORDER_TOTAL. Not all groups will be displayed, however. The HAVING clause displays only those groups whose SUM(NUM_ORDERED * QUOTED_PRICE) is greater than $1,000.

The order number, customer number, and order date are unique for each order. Thus, it would seem that merely grouping by order number would be sufficient. SQL requires that both the customer number and the order date be listed in the GROUP BY clause. Recall that a SELECT clause can include statistics calculated for only the groups or columns whose values are identical for each row in a group. By stating that the data is to be grouped by order number, customer number, and order date, you tell SQL that the values in these columns must be the same for each row in a group.

Using an Alias

When tables are listed in the FROM clause, you can give each table an **alias**, or an alternate name, that you can use in the rest of the statement. You create an alias by typing the name of the table, pressing the Spacebar, and then typing the name of the alias. No commas or periods are necessary to separate the two names.

One reason for using an alias is simplicity. In Example 8, you assign the REP table the alias R and the CUSTOMER table the alias C. By doing this, you can type R instead of REP and C instead of CUSTOMER in the remainder of the query. The query in this example is simple, so you might not see the full benefit of this feature. When a query is complex and requires you to qualify the names, using aliases can simplify the process.

EXAMPLE 8

List the number, last name, and first name for each sales rep together with the number and name for each customer the sales rep represents.

The query and its results using aliases appear in Figure 5-11.

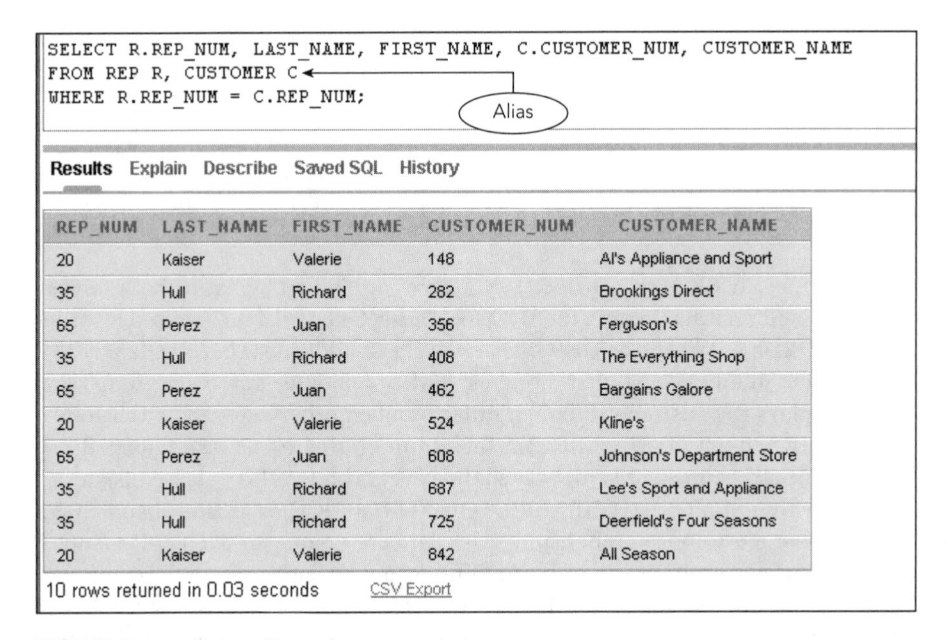

```
SELECT R.REP_NUM, LAST_NAME, FIRST_NAME, C.CUSTOMER_NUM, CUSTOMER_NAME
FROM REP R, CUSTOMER C
WHERE R.REP_NUM = C.REP_NUM;
```


REP_NUM	LAST_NAME	FIRST_NAME	CUSTOMER_NUM	CUSTOMER_NAME
20	Kaiser	Valerie	148	Al's Appliance and Sport
35	Hull	Richard	282	Brookings Direct
65	Perez	Juan	356	Ferguson's
35	Hull	Richard	408	The Everything Shop
65	Perez	Juan	462	Bargains Galore
20	Kaiser	Valerie	524	Kline's
65	Perez	Juan	608	Johnson's Department Store
35	Hull	Richard	687	Lee's Sport and Appliance
35	Hull	Richard	725	Deerfield's Four Seasons
20	Kaiser	Valerie	842	All Season

10 rows returned in 0.03 seconds CSV Export

FIGURE 5-11 Using aliases in a query

NOTE

Technically, it is unnecessary to qualify CUSTOMER_NUM because it is included only in the CUSTOMER table. It is qualified in Figure 5-11 for illustration purposes only.

Joining a Table to Itself

A second situation for using an alias is to join a table to itself, called a **self-join**, as illustrated in Example 9.

EXAMPLE 9

For each pair of customers located in the same city, display the customer number, customer name, and city.

If you had two separate tables for customers and the query requested customers in the first table having the same city as customers in the second table, you could use a normal join operation to find the answer. In this case, however, there is only *one* table (CUSTOMER) that stores all the customer information. You can treat the CUSTOMER table as if it were two tables in the query by creating an alias, as illustrated in Example 8. In this case, you use the following FROM clause:

```
FROM CUSTOMER F, CUSTOMER S
```

SQL treats this clause as a query of two tables: one that has the alias F (first), and another that has the alias S (second). The fact that both tables are really the same CUSTOMER table is not a problem. The query and its results appear in Figure 5-12.

```
SELECT F.CUSTOMER_NUM, F.CUSTOMER_NAME, S.CUSTOMER_NUM, S.CUSTOMER_NAME, F.CITY
FROM CUSTOMER F, CUSTOMER S
WHERE F.CITY = S.CITY
AND F.CUSTOMER_NUM < S.CUSTOMER_NUM
ORDER BY F.CUSTOMER_NUM, S.CUSTOMER_NUM;
```

Condition to join F and S tables

Results Explain Describe Saved SQL History

CUSTOMER_NUM	CUSTOMER_NAME	CUSTOMER_NUM	CUSTOMER_NAME	CITY
148	Al's Appliance and Sport	524	Kline's	Fillmore
282	Brookings Direct	462	Bargains Galore	Grove
282	Brookings Direct	842	All Season	Grove
462	Bargains Galore	842	All Season	Grove
608	Johnson's Department Store	725	Deerfield's Four Seasons	Sheldon

5 rows returned in 0.02 seconds CSV Export

FIGURE 5-12 Using aliases for a self-join

You are requesting a customer number and name from the F table, followed by a customer number and name from the S table, and then the city. (Because the city in the first table must match the city in the second table, you can select the city from either table.) The WHERE clause contains two conditions: the cities must match, and the customer number from the first table must be less than the customer number from the second table. In addition, the ORDER BY clause ensures that the data is sorted by the first customer number. For those rows with the same first customer number, the data is further sorted by the second customer number.

Q & A

Question: Why is the condition F.CUSTOMER_NUM < S.CUSTOMER_NUM important in the query?
Answer: If you did not include this condition, you would get the query results shown in Figure 5-13.

```
SELECT F.CUSTOMER_NUM, F.CUSTOMER_NAME, S.CUSTOMER_NUM, S.CUSTOMER_NAME, F.CITY
FROM CUSTOMER F, CUSTOMER S
WHERE F.CITY = S.CITY
ORDER BY F.CUSTOMER_NUM, S.CUSTOMER_NUM;
```

Results Explain Describe Saved SQL History

CUSTOMER_NUM	CUSTOMER_NAME	CUSTOMER_NUM	CUSTOMER_NAME	CITY
148	Al's Appliance and Sport	148	Al's Appliance and Sport	Fillmore
148	Al's Appliance and Sport	524	Kline's	Fillmore
282	Brookings Direct	282	Brookings Direct	Grove
282	Brookings Direct	462	Bargains Galore	Grove
282	Brookings Direct	842	All Season	Grove
356	Ferguson's	356	Ferguson's	Northfield
408	The Everything Shop	408	The Everything Shop	Crystal
462	Bargains Galore	282	Brookings Direct	Grove
462	Bargains Galore	462	Bargains Galore	Grove
462	Bargains Galore	842	All Season	Grove
524	Kline's	148	Al's Appliance and Sport	Fillmore
524	Kline's	524	Kline's	Fillmore
608	Johnson's Department Store	608	Johnson's Department Store	Sheldon
608	Johnson's Department Store	725	Deerfield's Four Seasons	Sheldon

FIGURE 5-13 Incorrect joining of a table to itself

The first row is included because it is true that customer number 148 (Al's Appliance and Sport) in the F table has the same city as customer number 148 (Al's Appliance and Sport) in the S table. The second row indicates that customer number 148 (Al's Appliance and Sport) has the same city as customer number 524 (Kline's). The eleventh row, however, repeats the same information because customer number 524 (Kline's) has the same city as customer number 148 (Al's Appliance and Sport). Of these three rows, the only row that should be included in the query results is the second row. The second row also is the only one of the three rows in which the first customer number (148) is less than the second customer number (524). This is why the query requires the condition F.CUSTOMER_NUM < S.CUSTOMER_NUM.

Using a Self-Join on a Primary Key Column

Figure 5-14 shows some fields from an EMPLOYEE table whose primary key is EMPLOYEE_NUM. Another field in the table is MGR_EMPLOYEE_NUM, which represents the number of the employee's manager, who also is an employee. If you look at the row for employee 206 (Joan Dykstra), you will see that employee 198 (Mona Canzler) is Joan's manager. By looking at the row for employee 198 (Mona Canzler), you see that her manager is employee 108 (Martin Holden). In the row for employee 108 (Martin Holden), the manager number is null, indicating that he has no manager.


```
SELECT EMPLOYEE_NUM, LAST_NAME, FIRST_NAME, MGR_EMPLOYEE_NUM
FROM EMPLOYEE;
```

Results Explain Describe Saved SQL History

EMPLOYEE_NUM	LAST_NAME	FIRST_NAME	MGR_EMPLOYEE_NUM
108	Holden	Martin	
198	Canzler	Mona	108
206	Dykstra	Joan	198
255	Murray	Steven	301
301	Galvez	Benito	108
366	Peterman	Beth	198
391	Traynor	Matt	301
402	Brent	Ashton	301
466	Scholten	Alyssa	108
551	Wiltzer	Morgan	198

10 rows returned in 0.02 seconds CSV Export

Employee 108 has no manager

Employee 198 manages employee 206

FIGURE 5-14 Employee and manager data

Suppose you need to list the employee number, employee last name, and employee first name along with the number, last name, and first name of each employee's manager. Just as in the previous self-join, you would list the EMPLOYEE table twice in the FROM clause with aliases.

The command shown in Figure 5-15 uses the letter E as an alias for the employee and the letter M as an alias for the manager. Thus E.EMPLOYEE_NUM is the employee's number and M.EMPLOYEE_NUM is the number of the employee's manager. In the SQL command, M.EMPLOYEE_NUM is renamed as MGR_NUM, M.LAST_NAME is renamed as MGR_LAST, and M.FIRST_NAME is renamed as MGR_FIRST. The condition in the WHERE clause ensures that E.MGR_EMPLOYEE_NUM (the number of the employee's manager) matches M.EMPLOYEE_NUM (the employee number on the manager's row in the table). Employee 108 is not included in the results because Martin Holden has no manager (see Figure 5-14).

```
SELECT E.EMPLOYEE_NUM, E.LAST_NAME, E.FIRST_NAME, M.EMPLOYEE_NUM AS MGR_NUM,
M.LAST_NAME AS MGR_LAST, M.FIRST_NAME AS MGR_FIRST
FROM EMPLOYEE E, EMPLOYEE M
WHERE E.MGR_EMPLOYEE_NUM = M.EMPLOYEE_NUM
ORDER BY E.EMPLOYEE_NUM;
```

Results Explain Describe Saved SQL History

EMPLOYEE_NUM	LAST_NAME	FIRST_NAME	MGR_NUM	MGR_LAST	MGR_FIRST
198	Canzler	Mona	108	Holden	Martin
206	Dykstra	Joan	198	Canzler	Mona
255	Murray	Steven	301	Galvez	Benito
301	Galvez	Benito	108	Holden	Martin
366	Peterman	Beth	198	Canzler	Mona
391	Traynor	Matt	301	Galvez	Benito
402	Brent	Ashton	301	Galvez	Benito
466	Scholten	Alyssa	108	Holden	Martin
551	Witzer	Morgan	198	Canzler	Mona

9 rows returned in 0.04 seconds CSV Export

FIGURE 5-15 List of employees and their managers

Joining Several Tables

It is possible to join several tables, as illustrated in Example 10. For each pair of tables you join, you must include a condition indicating how the columns are related.

EXAMPLE 10

For each part on order, list the part number, number ordered, order number, order date, customer number, and customer name, along with the last name of the sales rep who represents each customer.

A part is on order when it occurs on any row in the ORDER_LINE table. The part number, number ordered, and order number appear in the ORDER_LINE table. If these requirements represent the entire query, you would write the query as follows:

```
SELECT PART_NUM, NUM_ORDERED, ORDER_NUM
FROM ORDER_LINE;
```

This query is not sufficient, however. You also need the order date, which is in the ORDERS table; the customer number and name, which are in the CUSTOMER table; and the rep last name, which is in the REP table. Thus, you need to join *four* tables: ORDER_LINE, ORDERS, CUSTOMER, and REP. The procedure for joining more than two tables is essentially the same as the one for joining two tables. The difference is that the condition in the WHERE clause will be a compound condition. In this case, you would write the WHERE clause as follows:

```
WHERE ORDERS.ORDER_NUM = ORDER_LINE.ORDER_NUM
AND CUSTOMER.CUSTOMER_NUM = ORDERS.CUSTOMER_NUM
AND REP.REP_NUM = CUSTOMER.REP_NUM
```


The first condition relates an order to an order line with a matching order number. The second condition relates the customer to the order with a matching customer number. The final condition relates the rep to a customer with a matching sales rep number.

For the complete query, you list all the desired columns in the SELECT clause and qualify any columns that appear in more than one table. In the FROM clause, you list the tables that are involved in the query. The query and its results appear in Figure 5-16.

```
SELECT PART_NUM, NUM_ORDERED, ORDER_LINE.ORDER_NUM, ORDER_DATE, CUSTOMER.CUSTOMER_NUM,
CUSTOMER_NAME, LAST_NAME
FROM ORDER_LINE, ORDERS, CUSTOMER, REP
WHERE ORDERS.ORDER_NUM = ORDER_LINE.ORDER_NUM
AND CUSTOMER.CUSTOMER_NUM = ORDERS.CUSTOMER_NUM
AND REP.REP_NUM = CUSTOMER.REP_NUM;
```

Tables to include in query

Conditions to relate the tables

Results Explain Describe Saved SQL History

PART_NUM	NUM_ORDERED	ORDER_NUM	ORDER_DATE	CUSTOMER_NUM	CUSTOMER_NAME	LAST_NAME
AT94	11	21608	20-OCT-10	148	Al's Appliance and Sport	Kaiser
DR93	1	21610	20-OCT-10	356	Ferguson's	Perez
DW11	1	21610	20-OCT-10	356	Ferguson's	Perez
KL62	4	21613	21-OCT-10	408	The Everything Shop	Hull
KT03	2	21614	21-OCT-10	282	Brookings Direct	Hull
BV06	2	21617	23-OCT-10	608	Johnson's Department Store	Perez
CD52	4	21617	23-OCT-10	608	Johnson's Department Store	Perez
DR93	1	21619	23-OCT-10	148	Al's Appliance and Sport	Kaiser
KV29	2	21623	23-OCT-10	608	Johnson's Department Store	Perez

9 rows returned in 0.10 seconds CSV Export

FIGURE 5-16 Joining four tables in a query

Q & A

Question: Why is the PART_NUM column, which appears in the PART and ORDER_LINE tables, not qualified in the SELECT clause?

Answer: Among the tables listed in the query, only one table contains a column named PART_NUM, so it is not necessary to qualify the table. If the PART table also appeared in the FROM clause, you would need to qualify PART_NUM to avoid confusion between the PART_NUM columns in the PART and ORDER_LINE tables.

The query shown in Figure 5-16 is more complex than many of the previous ones you have examined. You might think that SQL is not such an easy language to use after all. If you take it one step at a time, however, the query in Example 10 really is not that difficult. To construct a detailed query in a step-by-step fashion, do the following:

1. List in the SELECT clause all the columns that you want to display. If the name of a column appears in more than one table, precede the column name with the table name (that is, qualify the column name).
2. List in the FROM clause all the tables involved in the query. Usually you include the tables that contain the columns listed in the SELECT clause. Occasionally, however, there might be a table that does not contain any columns used in the SELECT clause but that does contain columns used in the WHERE clause. In this case, you also must list the table in the FROM clause. For example, if you do not need to list a customer number or name, but you do need to list the rep name, you would not include any columns from the CUSTOMER table in the SELECT clause. The CUSTOMER table still is required, however, because you must include a column from it in the WHERE clause.
3. Take one pair of related tables at a time and indicate in the WHERE clause the condition that relates the tables. Join these conditions with the AND operator. If there are any other conditions, include them in the WHERE clause and connect them to the other conditions with the AND operator. For example, if you want to view parts present on orders placed by only those customers with $10,000 credit limits, you would add one more condition to the WHERE clause, as shown in Figure 5-17.

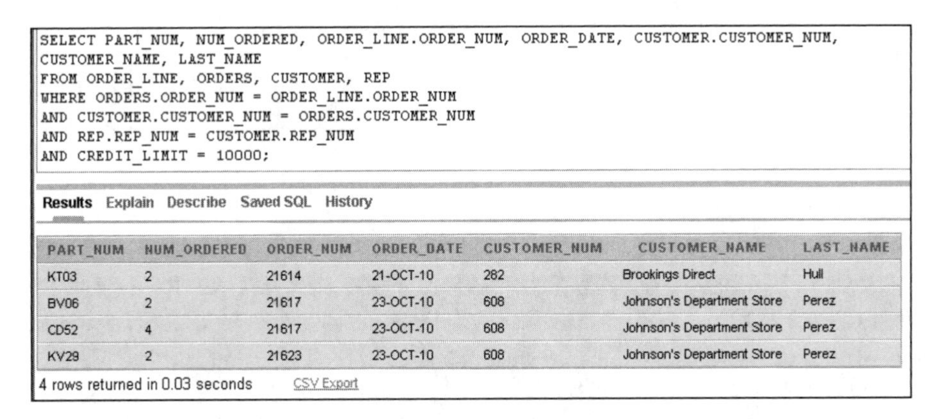

```
SELECT PART_NUM, NUM_ORDERED, ORDER_LINE.ORDER_NUM, ORDER_DATE, CUSTOMER.CUSTOMER_NUM,
CUSTOMER_NAME, LAST_NAME
FROM ORDER_LINE, ORDERS, CUSTOMER, REP
WHERE ORDERS.ORDER_NUM = ORDER_LINE.ORDER_NUM
AND CUSTOMER.CUSTOMER_NUM = ORDERS.CUSTOMER_NUM
AND REP.REP_NUM = CUSTOMER.REP_NUM
AND CREDIT_LIMIT = 10000;
```

Results Explain Describe Saved SQL History

PART_NUM	NUM_ORDERED	ORDER_NUM	ORDER_DATE	CUSTOMER_NUM	CUSTOMER_NAME	LAST_NAME
KT03	2	21614	21-OCT-10	282	Brookings Direct	Hull
BV06	2	21617	23-OCT-10	608	Johnson's Department Store	Perez
CD52	4	21617	23-OCT-10	608	Johnson's Department Store	Perez
KV29	2	21623	23-OCT-10	608	Johnson's Department Store	Perez

4 rows returned in 0.03 seconds CSV Export

FIGURE 5-17 Restricting the rows when joining four tables

SET OPERATIONS

In SQL, you can use the set operations for taking the union, intersection, and difference of two tables. The **union** of two tables uses the **UNION** operator to create a temporary table containing every row that is in either the first table, the second table, or both tables. The **intersection** of two tables uses the **INTERSECT** operator to create a temporary table containing all rows that are in both tables. The **difference** of two tables uses the **MINUS** operator to create a temporary table containing the set of all rows that are in the first table but that are not in the second table.

For example, suppose that TEMP1 is a table containing the number and name of each customer represented by sales rep 65. Further suppose that TEMP2 is a table containing the number and name of those customers that currently have orders on file, as shown in Figure 5-18.

TEMP1

CUSTOMER_NUM	CUSTOMER_NAME
356	Ferguson's
462	Bargains Galore
608	Johnson's Department Store

TEMP2

CUSTOMER_NUM	CUSTOMER_NAME
148	Al's Appliance and Sport
282	Brookings Direct
356	Ferguson's
408	The Everything Shop
608	Johnson's Department Store

FIGURE 5-18 Customers of rep 65 and customers with open orders

The union of TEMP1 and TEMP2 (TEMP1 UNION TEMP2) consists of the number and name of those customers that are represented by sales rep 65 *or* that currently have orders on file, *or* both. The intersection of these two tables (TEMP1 INTERSECT TEMP2) contains those customers that are represented by sales rep 65 *and* that have orders on file. The difference of these two tables (TEMP1 MINUS TEMP2) contains those customers that are represented by sales rep 65 but that *do not* have orders on file. The results of these set operations are shown in Figure 5-19.

TEMP1 UNION TEMP2

CUSTOMER_NUM	CUSTOMER_NAME
148	Al's Appliance and Sport
282	Brookings Direct
356	Ferguson's
408	The Everything Shop
462	Bargains Galore
608	Johnson's Department Store

TEMP1 INTERSECT TEMP2

CUSTOMER_NUM	CUSTOMER_NAME
356	Ferguson's
608	Johnson's Department Store

TEMP1 MINUS TEMP2

CUSTOMER_NUM	CUSTOMER_NAME
462	Bargains Galore

FIGURE 5-19 Union, intersection, and difference of the TEMP1 and TEMP2 tables

There is a restriction on set operations. It does not make sense, for example, to talk about the union of the CUSTOMER table and the ORDERS table because these tables do not contain the same columns. What might rows in this union look like? The two tables in the union *must* have the same structure for a union to be appropriate; the formal term is "union compatible." Two tables are **union compatible** when they have the same number of columns and their corresponding columns have identical data types and lengths.

Note that the definition of union compatible does not state that the columns of the two tables must be identical but rather that the columns must be of the same type. Thus, if one column is CHAR(20), the matching column also must be CHAR(20).

EXAMPLE 11

List the number and name of each customer that either is represented by sales rep 65 or that currently has orders on file, or both.

You can create a temporary table containing the number and name of each customer that is represented by sales rep 65 by selecting the customer numbers and names from the CUSTOMER table for which the sales rep number is 65. Then you can create another temporary table containing the number and name of each customer that currently has orders on file by joining the CUSTOMER and ORDERS tables. The two temporary tables created by this process have the same structure; that is, they both contain the CUSTOMER_NUM and CUSTOMER_NAME columns. Because the temporary tables are union compatible, it is possible to take the union of these two tables. The query and its results appear in Figure 5-20.

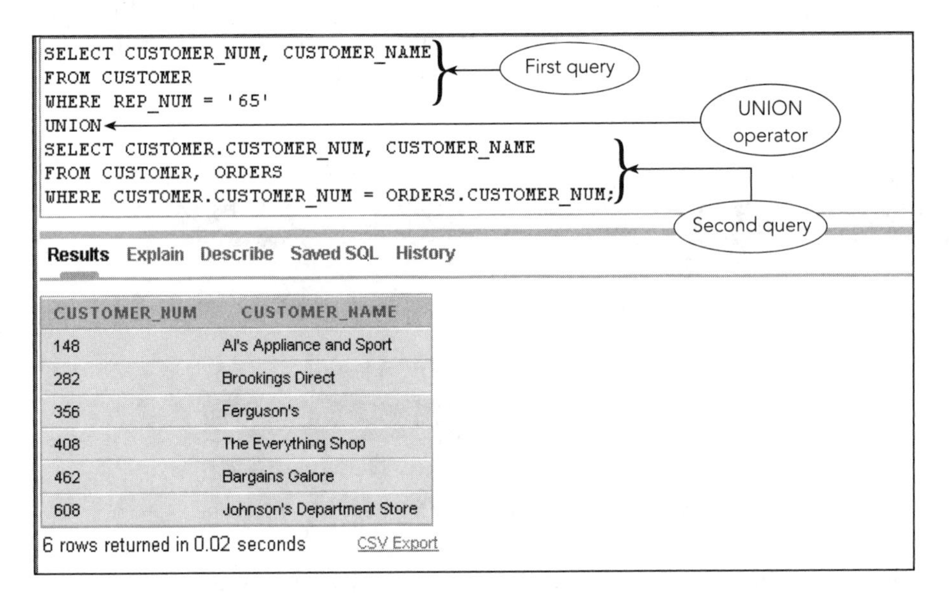

CUSTOMER_NUM	CUSTOMER_NAME
148	Al's Appliance and Sport
282	Brookings Direct
356	Ferguson's
408	The Everything Shop
462	Bargains Galore
608	Johnson's Department Store

FIGURE 5-20 Using the UNION operator

If your SQL implementation truly supports the union operation, it will remove any duplicate rows automatically. For example, any customer that is represented by sales rep 65 *and* that currently has orders on file will appear only once in the results. Oracle, Access, and SQL Server support the union operation and correctly remove duplicates.

EXAMPLE 12

List the number and name of each customer that is represented by sales rep 65 and that currently has orders on file.

The only difference between this query and the one in Example 11 is that the appropriate operator to use is INTERSECT, as shown in Figure 5-21.

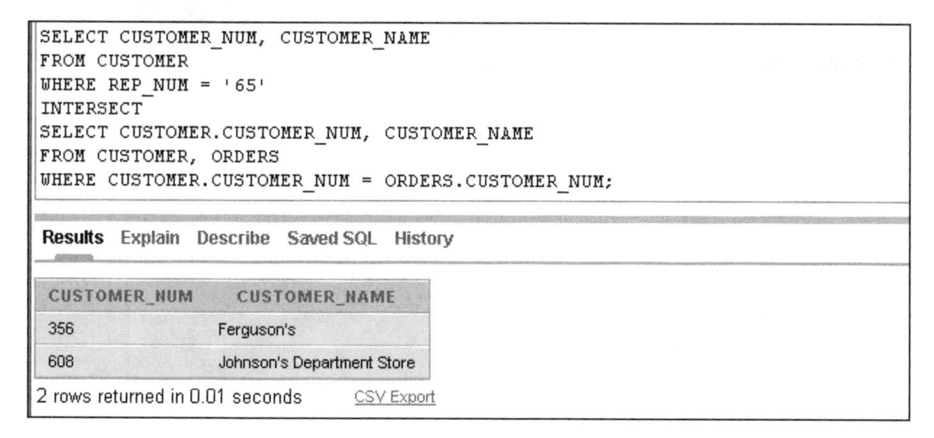

FIGURE 5-21 Using the INTERSECT operator

Some SQL implementations do not support the INTERSECT operator, so you need to take a different approach. The command shown in Figure 5-22 produces the same results as the INTERSECT operator by using the IN operator and a subquery. The command selects the number and name of each customer that is represented by sales rep 65 and whose customer number also appears in the collection of customer numbers in the ORDERS table.

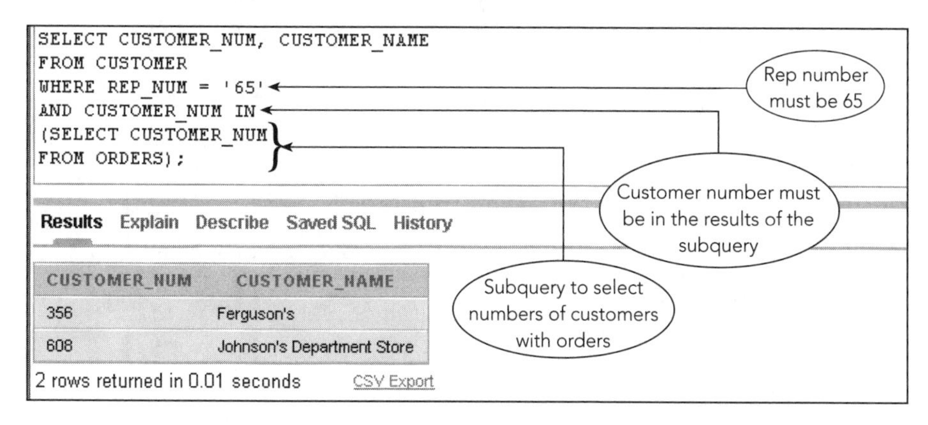

FIGURE 5-22 Performing an intersection without using the INTERSECT operator

NOTE

Oracle and SQL Server support the INTERSECT operator but Microsoft Access does not.

EXAMPLE 13

List the number and name of each customer that is represented by sales rep 65 but that does not have orders currently on file.

The query uses the MINUS operator, as shown in Figure 5-23.

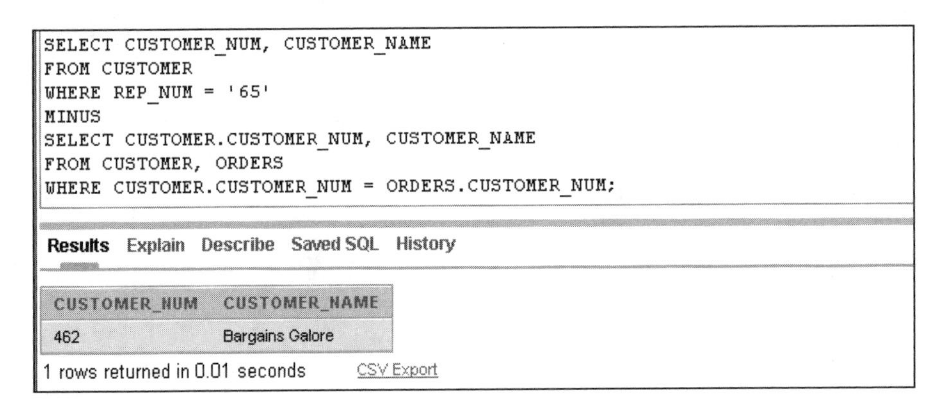

FIGURE 5-23 Using the MINUS operator

Just as with the INTERSECT operator, some SQL implementations do not support the MINUS operator. In such cases, you need to take a different approach, such as the one shown in Figure 5-24. This command produces the same results by selecting the number and name of each customer that is represented by sales rep 65 and whose customer number does *not* appear in the collection of customer numbers in the ORDERS table.

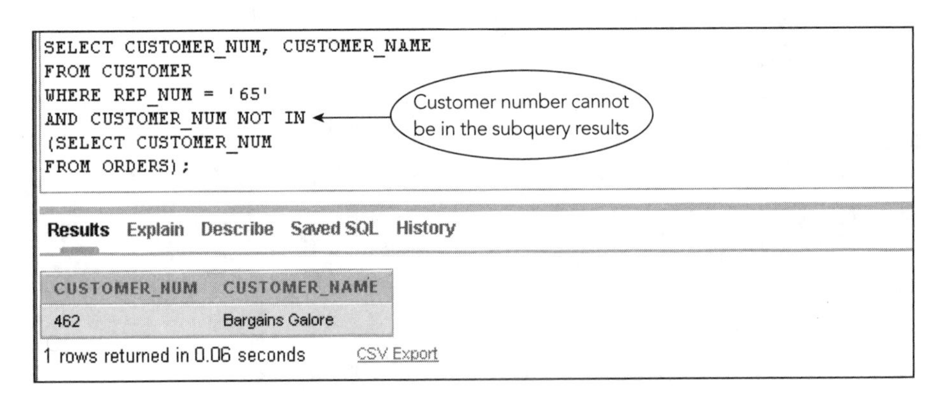

FIGURE 5-24 Performing a difference without using the MINUS operator

NOTE

Oracle supports the MINUS operator, but SQL Server and Microsoft Access do not.

ALL AND ANY

You can use the ALL and ANY operators with subqueries to produce a single column of numbers. When you precede the subquery by the **ALL** operator, the condition is true only if it satisfies *all* values produced by the subquery. When you precede the subquery by the **ANY** operator, the condition is true only if it satisfies *any* value (one or more) produced by the subquery. The following examples illustrate the use of these operators.

EXAMPLE 14

Find the customer number, name, current balance, and rep number of each customer whose balance exceeds the maximum balance of all customers represented by sales rep 65.

You can find the maximum balance of the customers represented by sales rep 65 in a subquery and then find all customers whose balances are greater than this number. There is an alternative method that is simpler, however. You can use the ALL operator, as shown in Figure 5-25.

```
SELECT CUSTOMER_NUM, CUSTOMER_NAME, BALANCE, REP_NUM
FROM CUSTOMER
WHERE BALANCE > ALL      <-- ALL operator
(SELECT BALANCE
FROM CUSTOMER
WHERE REP_NUM = '65');
```

Results Explain Describe Saved SQL History

CUSTOMER_NUM	CUSTOMER_NAME	BALANCE	REP_NUM
148	Al's Appliance and Sport	6550	20
524	Kline's	12762	20
842	All Season	8221	20

3 rows returned in 0.06 seconds CSV Export

FIGURE 5-25 SELECT command that uses the ALL operator

To some users, the query shown in Figure 5-25 might seem more natural than finding the maximum balance in the subquery. For other users, the opposite might be true. You can use whichever approach you prefer.

Q & A

Question: How would you get the same result for Example 14 without using the ALL operator?
Answer: You could select each customer whose balance is greater than the maximum balance of any customer of sales rep 65, as shown in Figure 5-26.


```
SELECT CUSTOMER_NUM, CUSTOMER_NAME, BALANCE, REP_NUM
FROM CUSTOMER
WHERE BALANCE >
(SELECT MAX(BALANCE)
FROM CUSTOMER
WHERE REP_NUM = '65');
```

Results Explain Describe Saved SQL History

CUSTOMER_NUM	CUSTOMER_NAME	BALANCE	REP_NUM
148	Al's Appliance and Sport	6550	20
524	Kline's	12762	20
842	All Season	8221	20

3 rows returned in 0.02 seconds CSV Export

FIGURE 5-26 Alternative to using the ALL operator

EXAMPLE 15

Find the customer number, name, current balance, and rep number of each customer whose balance is greater than the balance of at least one customer of sales rep 65.

You can find the minimum balance of the customers represented by sales rep 65 in a subquery and then find all customers whose balance is greater than this number. To simplify the process, you can use the ANY operator, as shown in Figure 5-27.

```
SELECT CUSTOMER_NUM, CUSTOMER_NAME, BALANCE, REP_NUM
FROM CUSTOMER
WHERE BALANCE > ANY
(SELECT BALANCE
FROM CUSTOMER
WHERE REP_NUM = '65');
```

ANY operator

Results Explain Describe Saved SQL History

CUSTOMER_NUM	CUSTOMER_NAME	BALANCE	REP_NUM
524	Kline's	12762	20
842	All Season	8221	20
148	Al's Appliance and Sport	6550	20
356	Ferguson's	5785	65
408	The Everything Shop	5285.25	35
462	Bargains Galore	3412	65
687	Lee's Sport and Appliance	2851	35

7 rows returned in 0.01 seconds CSV Export

FIGURE 5-27 SELECT command with an ANY operator

Q & A

Question: How would you get the same results without using the ANY operator?
Answer: You could select each customer whose balance is greater than the minimum balance of any customer of sales rep 65, as shown in Figure 5-28.

```
SELECT CUSTOMER_NUM, CUSTOMER_NAME, BALANCE, REP_NUM
FROM CUSTOMER
WHERE BALANCE >
(SELECT MIN(BALANCE)
FROM CUSTOMER
WHERE REP_NUM = '65');
```

Results Explain Describe Saved SQL History

CUSTOMER_NUM	CUSTOMER_NAME	BALANCE	REP_NUM
148	Al's Appliance and Sport	6550	20
356	Ferguson's	5785	65
408	The Everything Shop	5285.25	35
462	Bargains Galore	3412	65
524	Kline's	12762	20
687	Lee's Sport and Appliance	2851	35
842	All Season	8221	20

7 rows returned in 0.01 seconds CSV Export

FIGURE 5-28 Alternative to using the ANY operator

SPECIAL OPERATIONS

You can perform special operations within SQL, such as the self-join that you already used. Three other special operations are the inner join, the outer join, and the product.

Inner Join

A join that compares the tables in a FROM clause and lists only those rows that satisfy the condition in the WHERE clause is called an **inner join**. The joins that you have performed so far in this text have been inner joins. Example 16 illustrates the inner join.

EXAMPLE 16

Display the customer number, customer name, order number, and order date for each order. Sort the results by customer number.

This example requires the same type of join that you have been using. The command is:

```
SELECT CUSTOMER.CUSTOMER_NUM, CUSTOMER_NAME,
     ORDER_NUM, ORDER_DATE
FROM CUSTOMER, ORDERS
WHERE CUSTOMER.CUSTOMER_NUM = ORDERS.CUSTOMER_NUM
ORDER BY CUSTOMER.CUSTOMER_NUM;
```

The previous approach should work in any SQL implementation. An update to the SQL standard approved in 1992, called SQL-92, provides an alternative way of performing an inner join, as demonstrated in Figure 5-29.

```
SELECT CUSTOMER.CUSTOMER_NUM, CUSTOMER_NAME, ORDER_NUM, ORDER_DATE
FROM CUSTOMER
INNER JOIN ORDERS
ON CUSTOMER.CUSTOMER_NUM = ORDERS.CUSTOMER_NUM
ORDER BY CUSTOMER.CUSTOMER_NUM;
```

INNER JOIN clause

ON clause

Results Explain Describe Saved SQL History

CUSTOMER_NUM	CUSTOMER_NAME	ORDER_NUM	ORDER_DATE
148	Al's Appliance and Sport	21619	23-OCT-10
148	Al's Appliance and Sport	21608	20-OCT-10
282	Brookings Direct	21614	21-OCT-10
356	Ferguson's	21610	20-OCT-10
408	The Everything Shop	21613	21-OCT-10
608	Johnson's Department Store	21623	23-OCT-10
608	Johnson's Department Store	21617	23-OCT-10

7 rows returned in 0.01 seconds CSV Export

FIGURE 5-29 Query that uses an INNER JOIN clause

In the FROM clause, list the first table, and then include an INNER JOIN clause that includes the name of the second table. Instead of a WHERE clause, use an ON clause containing the same condition that you would have included in the WHERE clause.

Outer Join

Sometimes you need to list all the rows from one of the tables in a join, regardless of whether they match any rows in a second table. For example, you can perform the join of the CUSTOMER and ORDERS tables in the query for Example 16, but display all customers—even the ones without orders. This type of join is called an **outer join**.

There are actually three types of outer joins. In a **left outer join**, all rows from the table on the left (the table listed first in the query) are included regardless of whether they match rows from the table on the right (the table listed second in the query). Rows from the table on the right are included only when they match. In a **right outer join**, all rows from the table on the right are included regardless of whether they match rows from the table on the left. Rows from the table on the left are included only when they match. In a **full outer join**, all rows from both tables are included regardless of whether they match rows from the other table. (The full outer join is rarely used.)

Example 17 illustrates the use of a left outer join.

EXAMPLE 17

Display the customer number, customer name, order number, and order date for all orders. Include all customers in the results. For customers that do not have orders, omit the order number and order date.

To include all customers, you must perform an outer join. Assuming the CUSTOMER table is listed first, the join should be a left outer join. In SQL, you use the LEFT JOIN clause to perform a left outer join as shown in Figure 5-30. (You would use a RIGHT JOIN clause to perform a right outer join.)

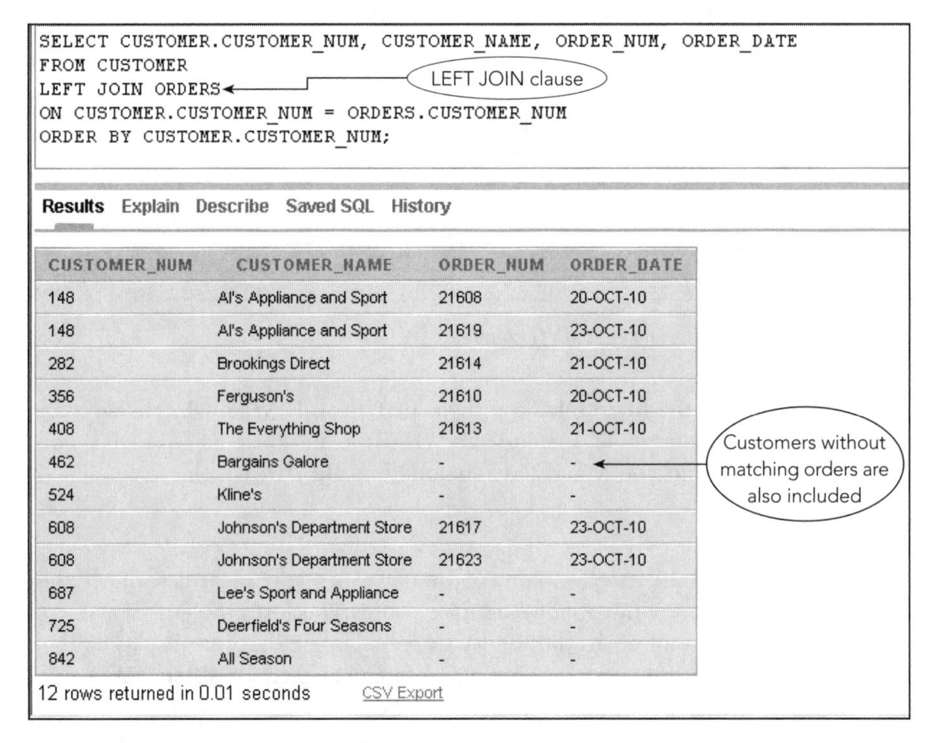

```
SELECT CUSTOMER.CUSTOMER_NUM, CUSTOMER_NAME, ORDER_NUM, ORDER_DATE
FROM CUSTOMER
LEFT JOIN ORDERS
ON CUSTOMER.CUSTOMER_NUM = ORDERS.CUSTOMER_NUM
ORDER BY CUSTOMER.CUSTOMER_NUM;
```


CUSTOMER_NUM	CUSTOMER_NAME	ORDER_NUM	ORDER_DATE
148	Al's Appliance and Sport	21608	20-OCT-10
148	Al's Appliance and Sport	21619	23-OCT-10
282	Brookings Direct	21614	21-OCT-10
356	Ferguson's	21610	20-OCT-10
408	The Everything Shop	21613	21-OCT-10
462	Bargains Galore	-	-
524	Kline's	-	-
608	Johnson's Department Store	21617	23-OCT-10
608	Johnson's Department Store	21623	23-OCT-10
687	Lee's Sport and Appliance	-	-
725	Deerfield's Four Seasons	-	-
842	All Season	-	-

12 rows returned in 0.01 seconds CSV Export

FIGURE 5-30 Query that uses a LEFT JOIN clause

All customers are included in the results. For customers without orders, the order number and date are blank. Technically, these blank values are null.

NOTE

In Oracle, there is another way to perform left and right outer joins. You write the join as you have been doing, with one exception. You include parentheses and a plus sign in the WHERE clause after the column in the table for which only matching rows are to be included. In this example, the plus sign would follow the CUSTOMER_NUM column in the ORDERS table because only orders that match customers are to be included. Because customers that do not have orders are to be included in the results, there would be no plus sign after the CUSTOMER_NUM column in the CUSTOMER table. The correct query is as follows:

```
SELECT CUSTOMER.CUSTOMER_NUM, CUSTOMER_NAME,
ORDER_NUM, ORDER_DATE
FROM CUSTOMER, ORDERS
WHERE CUSTOMER.CUSTOMER_NUM = ORDERS.CUSTOMER_NUM(+)
ORDER BY CUSTOMER.CUSTOMER_NUM;
```

Running this query produces the same results shown in Figure 5-30.

Product

The **product** (formally called the **Cartesian product**) of two tables is the combination of all rows in the first table and all rows in the second table.

NOTE

The product operation is not common. You need to be aware of it, however, because it is easy to create a product inadvertently by omitting the WHERE clause when you are attempting to join tables.

EXAMPLE 18

Form the product of the CUSTOMER and ORDERS tables. Display the customer number and name from the CUSTOMER table, along with the order number and order date from the ORDERS table.

Forming a product is actually very easy. You simply omit the WHERE clause, as shown in Figure 5-31.

```
SELECT CUSTOMER.CUSTOMER_NUM, CUSTOMER_NAME, ORDER_NUM, ORDER_DATE
FROM CUSTOMER, ORDERS;
```

No condition relates the tables in the FROM clause

Results Explain Describe Saved SQL History

CUSTOMER_NUM	CUSTOMER_NAME	ORDER_NUM	ORDER_DATE
148	Al's Appliance and Sport	21608	20-OCT-10
282	Brookings Direct	21608	20-OCT-10
356	Ferguson's	21608	20-OCT-10
408	The Everything Shop	21608	20-OCT-10
462	Bargains Galore	21608	20-OCT-10
524	Kline's	21608	20-OCT-10
608	Johnson's Department Store	21608	20-OCT-10
687	Lee's Sport and Appliance	21608	20-OCT-10
725	Deerfield's Four Seasons	21608	20-OCT-10
842	All Season	21608	20-OCT-10
148	Al's Appliance and Sport	21610	20-OCT-10
282	Brookings Direct	21610	20-OCT-10
356	Ferguson's	21610	20-OCT-10
408	The Everything Shop	21610	20-OCT-10
462	Bargains Galore	21610	20-OCT-10

FIGURE 5-31 Query that produces a product of two tables

Q & A

Question: Figure 5-31 does not show all the rows in the result. How many rows are actually included?

Answer: The CUSTOMER table has 10 rows and the ORDERS table has seven rows. Because each of the 10 customer rows is matched with each of the seven order rows, there are 70 (10 x 7) rows in the result.

Chapter Summary

- To join tables, indicate in the SELECT clause all columns to display, list in the FROM clause all tables to join, and then include in the WHERE clause any conditions requiring values in matching columns to be equal.
- When referring to matching columns in different tables, you must qualify the column names to avoid confusion. You qualify column names using the following format: table name.column name.
- Use the IN or EXISTS operators with an appropriate subquery as an alternate way of performing a join.
- A subquery can contain another subquery. The innermost subquery is executed first.
- The name of a table in a FROM clause can be followed by an alias, which is an alternate name for a table. You can use the alias in place of the table name throughout the SQL command. By using two different aliases for the same table in a single SQL command, you can join a table to itself.
- The UNION operator creates a union of two tables (the collection of rows that are in either or both tables). The INTERSECT operator creates the intersection of two tables (the collection of rows that are in both tables). The MINUS operator creates the difference of two tables (the collection of rows that are in the first table but not in the second table). To perform any of these operations, the tables involved must be union compatible. Two tables are union compatible when they have the same number of columns and their corresponding columns have identical data types and lengths.
- When the ALL operator precedes a subquery, the condition is true only if it is satisfied by *all* values produced by the subquery.
- When the ANY operator precedes a subquery, the condition is true only if it is satisfied by *any* value (one or more) produced by the subquery.
- In an inner join, only matching rows from both tables are included. You can use the INNER JOIN clause to perform an inner join.
- In a left outer join, all rows from the table on the left (the table listed first in the query) are included regardless of whether they match rows from the table on the right (the table listed second in the query). Rows from the table on the right are included only when they match. You can use the LEFT JOIN clause to perform a left outer join. In a right outer join, all rows from the table on the right are included regardless of whether they match rows from the table on the left. Rows from the table on the left are included only when they match. You can use the RIGHT JOIN clause to perform a right outer join.
- The product (Cartesian product) of two tables is the combination of all rows in the first table and all rows in the second table. To form a product of two tables, include both tables in the FROM clause and omit the WHERE clause.

Key Terms

alias
ALL
ANY
Cartesian product
correlated subquery
difference
EXISTS
full outer join
inner join
INTERSECT
intersection
join
left outer join
MINUS
nested subquery
outer join
product
right outer join
self-join
union
UNION
union compatible

Review Questions

1. How do you join tables in SQL?
2. When must you qualify names in SQL commands? How do you qualify a column name?
3. List two operators that you can use with subqueries as an alternate way of performing joins.
4. What is a nested subquery? In which order does SQL evaluate nested subqueries?
5. What is an alias? How do you specify an alias in SQL? How do you use an alias?
6. How do you join a table to itself in SQL?
7. How do you take the union of two tables in SQL? How do you take the intersection of two tables in SQL? How do you take the difference of two tables in SQL? Are there any restrictions on the tables when performing any of these operations?
8. What does it mean for two tables to be union compatible?
9. How do you use the ALL operator with a subquery?
10. How do you use the ANY operator with a subquery?
11. Which rows are included in an inner join? What clause can you use to perform an inner join in SQL?
12. Which rows are included in a left outer join? What clause can you use to perform a left outer join in SQL?
13. Which rows are included in a right outer join? What clause can you use to perform a right outer join in SQL?
14. What is the formal name for the product of two tables? How do you form a product in SQL?
15. Use your favorite Web browser and Web search engine to find definitions for the terms equi-join, natural join, and cross join. Write a short report that identifies how these terms relate to the terms join, inner join, and Cartesian product. Be sure to reference your online sources properly.

16. Use your favorite Web browser and Web search engine to find information on cost-based query optimizers. Write a short report that explains how cost-based query optimization works, and what type(s) of queries benefit the most from cost-based query optimization. Be sure to reference your online sources properly.

Exercises

Premiere Products

Use SQL and the Premiere Products database (see Figure 1-2 in Chapter 1) to complete the following exercises. If directed to do so by your instructor, use the information provided with the Chapter 3 Exercises to print your output.

1. For each order, list the order number and order date along with the number and name of the customer that placed the order.
2. For each order placed on October 23, 2010, list the order number along with the number and name of the customer that placed the order.
3. For each order, list the order number, order date, part number, number of units ordered, and quoted price for each order line that makes up the order.
4. Use the IN operator to find the number and name of each customer that placed an order on October 23, 2010.
5. Repeat Exercise 4, but this time use the EXISTS operator in your answer.
6. Find the number and name of each customer that did not place an order on October 23, 2010.
7. For each order, list the order number, order date, part number, part description, and item class for each part that makes up the order.
8. Repeat Exercise 7, but this time order the rows by item class and then by order number.
9. Use a subquery to find the rep number, last name, and first name of each sales rep who represents at least one customer with a credit limit of $10,000. List each sales rep only once in the results.
10. Repeat Exercise 9, but this time do not use a subquery.
11. Find the number and name of each customer that currently has an order on file for a Gas Range.
12. List the part number, part description, and item class for each pair of parts that are in the same item class. (For example, one such pair would be part AT94 and part FD21, because the item class for both parts is HW.)
13. List the order number and order date for each order placed by the customer named Johnson's Department Store. (*Hint:* To enter an apostrophe (single quotation mark) within a string of characters, type two single quotation marks.)
14. List the order number and order date for each order that contains an order line for an Iron.
15. List the order number and order date for each order that either was placed by Johnson's Department Store or that contains an order line for a Gas Range.
16. List the order number and order date for each order that was placed by Johnson's Department Store and that contains an order line for a Gas Range.

17. List the order number and order date for each order that was placed by Johnson's Department Store but that does not contain an order line for a Gas Range.
18. List the part number, part description, unit price, and item class for each part that has a unit price greater than the unit price of every part in item class AP. Use either the ALL or ANY operator in your query. (*Hint:* Make sure you select the correct operator.)
19. If you used ALL in Exercise 18, repeat the exercise using ANY. If you used ANY, repeat the exercise using ALL, and then run the new command. What question does this command answer?
20. For each part, list the part number, description, units on hand, order number, and number of units ordered. All parts should be included in the results. For those parts that are currently not on order, the order number and number of units ordered should be left blank. Order the results by part number.

Henry Books

Use SQL and the Henry Books database (see Figures 1-4 through 1-7 in Chapter 1) to complete the following exercises. If directed to do so by your instructor, use the information provided with the Chapter 3 Exercises to print your output.

1. For each book, list the book code, book title, publisher code, and publisher name. Order the results by publisher name.
2. For each book published by Scribner, list the book code, book title, and price.
3. List the book title, book code, and price of each book published by Scribner that has a book price of at least $14.
4. List the book code, book title, and units on hand for each book in branch number 3.
5. List the book title for each book that has the type PSY and that is published by Berkley Publishing.
6. Find the book title for each book written by author number 18. Use the IN operator in your query.
7. Repeat Exercise 6, but this time use the EXISTS operator in your query.
8. Find the book code and book title for each book located in branch number 2 and written by author 20.
9. List the book codes for each pair of books that have the same price. (For example, one such pair would be book 0200 and book 7559, because the price of both books is $8.00.) The first book code listed should be the major sort key, and the second book code should be the minor sort key.
10. Find the book title, author last name, and units on hand for each book in branch number 4.
11. Repeat Exercise 10, but this time list only paperback books.
12. Find the book code and book title for each book whose price is more than $10 or that was published in Boston.
13. Find the book code and book title for each book whose price is more than $10 and that was published in Boston.
14. Find the book code and book title for each book whose price is more than $10 but that was not published in Boston.

15. Find the book code and book title for each book whose price is greater than the book price of every book that has the type MYS.
16. Find the book code and book title for each book whose price is greater than the price of at least one book that has the type MYS.
17. List the book code, book title, and units on hand for each book in branch number 2. Be sure each book is included, regardless of whether there are any copies of the book currently on hand in branch 2. Order the output by book code.

Alexamara Marina Group

Use SQL and the Alexamara Marina Group database (see Figures 1-8 through 1-12 in Chapter 1) to complete the following exercises. If directed to do so by your instructor, use the information provided with the Chapter 3 Exercises to print your output.

1. For every boat, list the marina number, slip number, boat name, owner number, owner's first name, and owner's last name.
2. For every completed or open service request for routine engine maintenance, list the slip ID, description, and status.
3. For every service request for routine engine maintenance, list the slip ID, marina number, slip number, estimated hours, spent hours, owner number, and owner's last name.
4. List the first and last names of all owners who have a boat in a 30-foot slip. Use the IN operator in your query.
5. Repeat Exercise 4, but this time use the EXISTS operator in your query.
6. List the names of any pair of boats that have the same type. For example, one pair would be *Anderson II* and *Escape*, because the boat type for both boats is Sprite 4000. The first name listed should be the major sort key and the second name should be the minor sort key.
7. List the boat name, owner number, owner last name, and owner first name for each boat in marina 1.
8. Repeat Exercise 7, but this time only list boats in 40-foot slips.
9. List the marina number, slip number, and boat name for boats whose owners live in Glander Bay or whose type is Sprite 4000.
10. List the marina number, slip number, and boat name for boats whose owners live in Glander Bay and whose type is Sprite 4000.
11. List the marina number, slip number, and boat name for boats whose owners live in Glander Bay but whose type is not Sprite 4000.
12. Find the service ID and slip ID for each service request whose estimated hours is greater than the number of estimated hours of at least one service request on which the category number is 3.
13. Find the service ID and slip ID for each service request whose estimated hours is greater than the number of estimated hours on every service request on which the category number is 3.
14. List the slip ID, boat name, owner number, service ID, number of estimated hours, and number of spent hours for each service request on which the category number is 2.
15. Repeat Exercise 14, but this time be sure each slip is included regardless of whether the boat in the slip currently has any service requests for category 2.

CHAPTER 6

UPDATING DATA

LEARNING OBJECTIVES

Objectives

- Create a new table from an existing table
- Change data using the UPDATE command
- Add new data using the INSERT command
- Delete data using the DELETE command
- Use nulls in an UPDATE command
- Change the structure of an existing table
- Use the COMMIT and ROLLBACK commands to make permanent data updates or to reverse updates
- Understand transactions and the role of COMMIT and ROLLBACK in supporting transactions
- Drop a table

INTRODUCTION

In this chapter, you will learn how to create a new table from an existing table and make changes to the data in a table. You will use the UPDATE command to change data in one or more rows in a table, and use the INSERT command to add new rows. You will use the DELETE command to delete rows. You will learn how to change the structure of a table in a variety of ways and use nulls in update operations. You will use the COMMIT command to make changes permanent and use the ROLLBACK command to undo changes, and understand how to use these commands in transactions. Finally, you will learn how to delete a table and its data.

CREATING A NEW TABLE FROM AN EXISTING TABLE

You can create a new table using data in an existing table, as illustrated in the following examples.

EXAMPLE 1

Create a new table named LEVEL1_CUSTOMER that contains the following columns from the CUSTOMER table: CUSTOMER_NUM, CUSTOMER_NAME, BALANCE, CREDIT_LIMIT, and REP_NUM. The columns in the new LEVEL1_CUSTOMER table should have the same characteristics as the corresponding columns in the CUSTOMER table.

You describe the new table named LEVEL1_CUSTOMER by using the CREATE TABLE command shown in Figure 6-1.

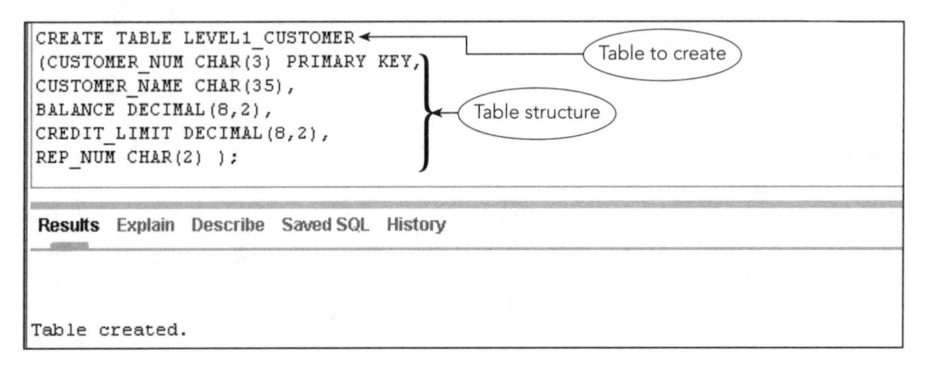

FIGURE 6-1 Creating the LEVEL1_CUSTOMER table

ACCESS USER NOTE

If you are using Access to create the LEVEL1_CUSTOMER table, use the CURRENCY data type instead of the DECIMAL data type for the BALANCE and CREDIT_LIMIT fields. (Access does not support the DECIMAL data type.) You do not need to enter the field size and number of decimal places when using the CURRENCY data type.

EXAMPLE 2

Insert into the LEVEL1_CUSTOMER table the customer number, customer name, balance, credit limit, and rep number for customers with credit limits of $7,500.

You can create a SELECT command to select the desired data from the CUSTOMER table, just as you did in Chapter 4. By placing this SELECT command in an INSERT command, you can add the query results to a table. The INSERT command appears in Figure 6-2; this command inserts four rows into the LEVEL1_CUSTOMER table.

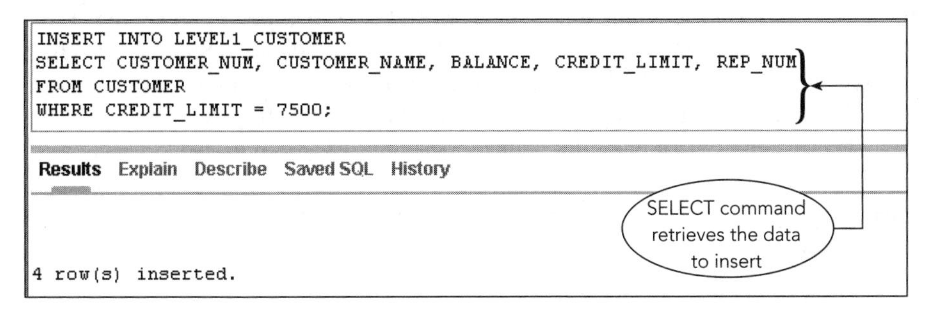

FIGURE 6-2 INSERT command to add data to the LEVEL1_CUSTOMER table

The SELECT command shown in Figure 6-3 displays the data in the LEVEL1_CUSTOMER table. Notice that the data comes from the new table you just created (LEVEL1_CUSTOMER), and not from the CUSTOMER table.

```
SELECT *
FROM LEVEL1_CUSTOMER;
```

Results Explain Describe Saved SQL History

CUSTOMER_NUM	CUSTOMER_NAME	BALANCE	CREDIT_LIMIT	REP_NUM
148	Al's Appliance and Sport	6550	7500	20
356	Ferguson's	5785	7500	65
725	Deerfield's Four Seasons	248	7500	35
842	All Season	8221	7500	20

4 rows returned in 0.15 seconds CSV Export

FIGURE 6-3 LEVEL1_CUSTOMER data

CHANGING EXISTING DATA IN A TABLE

The data stored in tables is subject to constant change; prices, addresses, commission amounts, and other data in a database change on a regular basis. To keep data current, you must be able to make these changes to the data in your tables. You can use the **UPDATE** command to change rows for which a specific condition is true.

EXAMPLE 3

Change the name of customer 842 in the LEVEL1_CUSTOMER table to "All Season Sport."

The format for the UPDATE command is the word UPDATE, followed by the name of the table to be updated. The next portion of the command consists of the word SET, followed by the name of the column to be updated, an equals sign, and the new value. When necessary, include a WHERE clause to indicate the row(s) on which the change is to occur. The UPDATE command shown in Figure 6-4 changes the name of customer 842 to All Season Sport.

```
UPDATE LEVEL1_CUSTOMER
SET CUSTOMER_NAME = 'All Season Sport'
WHERE CUSTOMER_NUM = '842';
```

Results Explain Describe Saved SQL History

1 row(s) updated.

FIGURE 6-4 UPDATE command to change the name of customer 842

The SELECT command shown in Figure 6-5 shows the data in the table after the change has been made. It is a good idea to use a SELECT command to display the data you changed to verify that the correct update was made.

```
SELECT *
FROM LEVEL1_CUSTOMER;
```

Results Explain Describe Saved SQL History

CUSTOMER_NUM	CUSTOMER_NAME	BALANCE	CREDIT_LIMIT	REP_NUM
148	Al's Appliance and Sport	6550	7500	20
356	Ferguson's	5785	7500	65
725	Deerfield's Four Seasons	248	7500	35
842	All Season Sport	8221	7500	20

4 rows returned in 0.06 seconds CSV Export

FIGURE 6-5 LEVEL1_CUSTOMER table after update

EXAMPLE 4

For each customer in the LEVEL1_CUSTOMER table that is represented by sales rep 20 and also has a balance that does not exceed the credit limit, increase the customer's credit limit to $8,000.

The only difference between Examples 3 and 4 is that Example 4 uses a compound condition to identify the row(s) to be changed. The UPDATE command appears in Figure 6-6.


```
UPDATE LEVEL1_CUSTOMER
SET CREDIT_LIMIT = 8000
WHERE REP_NUM = '20'
AND BALANCE < CREDIT_LIMIT;
```

Results Explain Describe Saved SQL History

1 row(s) updated.

FIGURE 6-6 Using a compound condition in an update

The SELECT command shown in Figure 6-7 shows the table after the update.

```
SELECT *
FROM LEVEL1_CUSTOMER;
```

Results Explain Describe Saved SQL History

CUSTOMER_NUM	CUSTOMER_NAME	BALANCE	CREDIT_LIMIT	REP_NUM
148	Al's Appliance and Sport	6550	8000	20
356	Ferguson's	5785	7500	65
725	Deerfield's Four Seasons	248	7500	35
842	All Season Sport	8221	7500	20

4 rows returned in 0.08 seconds CSV Export

FIGURE 6-7 Credit limit increased for customer number 148

You also can use the existing value in a column and a calculation to update a value. For example, when you need to increase the credit limit by 10 percent instead of changing it to a specific value, you can multiply the existing credit limit by 1.10. The following SET clause makes this change:

```
SET CREDIT_LIMIT = CREDIT_LIMIT * 1.10
```

ADDING NEW ROWS TO AN EXISTING TABLE

In Chapter 3, you used the INSERT command to add the initial rows to the tables in the database. You also can use the INSERT command to add additional rows to tables.

EXAMPLE 5

Add customer number 895 to the LEVEL1_CUSTOMER table. The name is Peter and Margaret's, the balance is 0, the credit limit is $8,000, and the rep number is 20.

The appropriate INSERT command is shown in Figure 6-8. Because the name "Peter and Margaret's" contains an apostrophe, you type two single quotation marks to create the apostrophe.

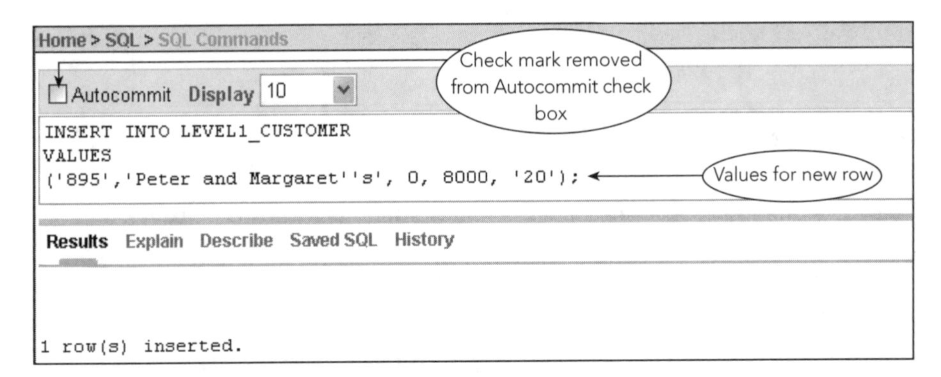

FIGURE 6-8 Inserting a row

The SELECT command in Figure 6-9 shows that the row was successfully added.

```
SELECT *
FROM LEVEL1_CUSTOMER;
```

Results Explain Describe Saved SQL History

CUSTOMER_NUM	CUSTOMER_NAME	BALANCE	CREDIT_LIMIT	REP_NUM
148	Al's Appliance and Sport	6550	8000	20
356	Ferguson's	5785	7500	65
725	Deerfield's Four Seasons	248	7500	35
842	All Season Sport	8221	7500	20
895	Peter and Margaret's	0	8000	20

5 rows returned in 0.12 seconds

FIGURE 6-9 Customer 895 added to LEVEL1_CUSTOMER table

NOTE

Your output might be sorted in a different order from what is shown in Figure 6-9. If you need to sort the rows in a specific order, use an ORDER BY clause with the desired sort key(s).

COMMIT AND ROLLBACK

Figure 6-8 shows that the user cleared the check mark from the Autocommit check box before running the query. **Autocommit** is the default transaction mode and commits (makes permanent) each action query (INSERT, UPDATE, DELETE) as soon as the user executes the query. Although the Autocommit transaction mode is fine for most action queries, there are times when the user needs better control over when a transaction is committed. This is particularly important in multi-user database applications when more than one person can update the database and in applications when users are running script files that contain multiple updates. When you need more control over when transactions are committed, you should disable the Autocommit feature by clearing its check box before executing a query.

If you do not use Autocommit, queries that include updates to table data are only temporary and you can reverse (cancel) them at any time during your current work session. Updates become permanent automatically when you exit from the DBMS. If you are not using Autocommit during your current work session, however, you can still **commit** (save) your changes immediately by executing the **COMMIT** command.

If you decide that you do not want to save the changes you have made during your current work session, you can **roll back** (reverse) the changes by executing the **ROLLBACK** command. Any updates made since you ran the most recent COMMIT command will be reversed when you run the ROLLBACK command. If you have not run the COMMIT command, executing the ROLLBACK command will reverse all updates made during the current work session. You should note that the ROLLBACK command reverses only changes made to the data; it does not reverse changes made to a table's structure. For example, if you change the length of a character column, you cannot use the ROLLBACK command to return the column length to its original state.

If you determine that an update was made incorrectly, you can use the ROLLBACK command to return the data to its original state. If, on the other hand, you have verified that the update you made is correct, you can use the COMMIT command to make the update permanent. You do this by typing COMMIT; after running the update. However, you should note that the COMMIT command is permanent; after executing a COMMIT command, running the ROLLBACK command cannot reverse the update.

ACCESS USER NOTE

Access does not support the COMMIT or ROLLBACK commands.

SQL SERVER USER NOTE

In SQL Server, the commands used to commit and roll back data are COMMIT TRANSACTION and ROLLBACK TRANSACTION. By default, SQL Server is in Autocommit transaction mode. To turn off the Autocommit feature, execute the following command:

```
SET XACT_ABORT ON
```

To turn the Autocommit feature back on, execute the following command:

```
SET XACT_ABORT OFF
```


TRANSACTIONS

A **transaction** is a logical unit of work. You can think of a transaction as a sequence of steps that accomplish a single task. When discussing transactions, it is essential that the entire sequence is completed successfully.

For example, to enter an order, you must add the corresponding order to the ORDERS table, and then add each order line in the order to the ORDER_LINE table. These multiple steps accomplish the "single" task of entering an order. Suppose you have added the order and the first order line, but you are unable to enter the second order line for some reason; perhaps the part on the order line does not exist. This problem would leave the order in a partially entered state, which is unacceptable. To prevent this problem, you would execute a rollback, thus reversing the insertion of the order and the first order line.

You can use the COMMIT and ROLLBACK commands to support transactions as follows:

- Before beginning the updates for a transaction, commit any previous updates by executing the COMMIT command.
- Complete the updates for the transaction. If any update cannot be completed, execute the ROLLBACK command and discontinue the updates for the current transaction.
- If you can complete all updates successfully, execute the COMMIT command after completing the final update.

CHANGING AND DELETING EXISTING ROWS

As you learned in Chapter 3, you use the DELETE command to remove rows from a table. In Example 6, you will change data and then use the DELETE command to delete a customer from the LEVEL1_CUSTOMER table. In Example 7, you will execute a rollback to reverse the updates made in Example 6. In this case, the rollback will return the row to its previous state and reinstate the deleted record.

EXAMPLE 6

In the LEVEL1_CUSTOMER table, change the name of customer 356 to "Smith Sport," and then delete customer 895.

To delete data from the database, use the DELETE command. The format for the **DELETE** command is the word DELETE followed by the name of the table containing the row(s) to be deleted. Next, use a WHERE clause with a condition to select the row(s) to delete. All rows satisfying the condition will be deleted.

The first part of Example 6 requests a name change for customer 356; the command shown in Figure 6-10 makes this change.


```
UPDATE LEVEL1_CUSTOMER
SET CUSTOMER_NAME = 'Smith Sport'
WHERE CUSTOMER_NUM = '356';
```

Results Explain Describe Saved SQL History

```
1 row(s) updated.
```

FIGURE 6-10 Using an UPDATE command to change the name of customer 356

The second part of Example 6 requires deleting customer 895; this command is shown in Figure 6-11.

```
DELETE FROM LEVEL1_CUSTOMER
WHERE CUSTOMER_NUM = '895';
```

Results Explain Describe Saved SQL History

```
1 row(s) deleted.
```

FIGURE 6-11 Using a DELETE command to delete customer 895

The command shown in Figure 6-12 displays the data in the table, verifying the change and the deletion.

```
SELECT *
FROM LEVEL1_CUSTOMER;
```

Results Explain Describe Saved SQL History

CUSTOMER_NUM	CUSTOMER_NAME	BALANCE	CREDIT_LIMIT	REP_NUM
148	Al's Appliance and Sport	6550	8000	20
356	Smith Sport	5785	7500	65
725	Deerfield's Four Seasons	248	7500	35
842	All Season Sport	8221	7500	20

4 rows returned in 0.06 seconds CSV Export

FIGURE 6-12 Results of update and delete

Q & A

Question: What happens when you run a DELETE command that does not contain a WHERE clause?
Answer: Without a condition to specify which row(s) to delete, the query will delete all rows from the table.

Executing a Rollback

The following example executes a rollback.

EXAMPLE 7

Execute a rollback and then display the data in the LEVEL1_CUSTOMER table.

To execute a rollback, execute the ROLLBACK command, as shown in Figure 6-13.

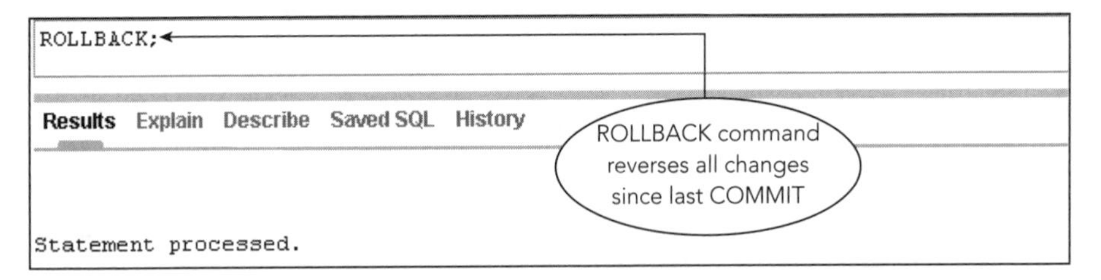

FIGURE 6-13 Executing a rollback

Figure 6-14 shows a SELECT command for the LEVEL1_CUSTOMER table after executing the rollback. Notice that the name of customer 356 has changed back to Ferguson's and the row for customer 895 has been reinstated. All updates made prior to the previous commit are still reflected in the data.

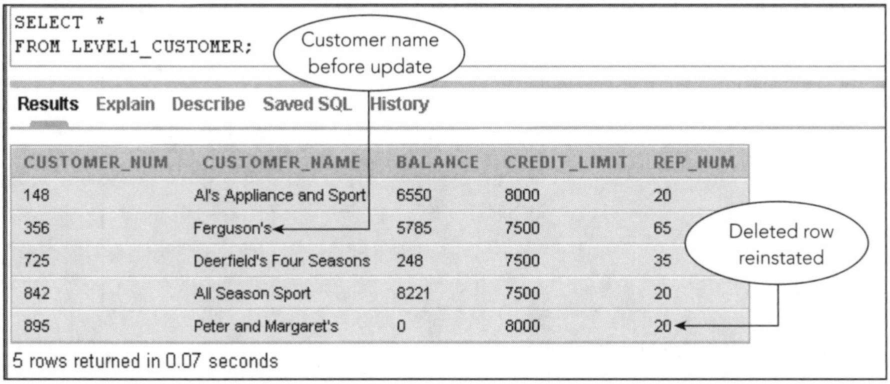

```
SELECT *
FROM LEVEL1_CUSTOMER;
```

Results Explain Describe Saved SQL History

CUSTOMER_NUM	CUSTOMER_NAME	BALANCE	CREDIT_LIMIT	REP_NUM
148	Al's Appliance and Sport	6550	8000	20
356	Ferguson's	5785	7500	65
725	Deerfield's Four Seasons	248	7500	35
842	All Season Sport	8221	7500	20
895	Peter and Margaret's	0	8000	20

5 rows returned in 0.07 seconds

FIGURE 6-14 Data in the LEVEL1_CUSTOMER table after executing a rollback

ACCESS USER NOTE

If you are using Access to complete these steps, you will not be able to execute the ROLLBACK command. Consequently, your data for the remaining examples in this chapter will differ slightly from the data shown in the figures—customer 356 will be named Smith Sport and customer 895 will not be included.

NOTE

In the remaining examples in this chapter, the Autocommit feature is enabled, that is, there is a check mark in the Autocommit check box. All updates are committed immediately without requiring any special action on your part. In addition, it will no longer be possible to roll back updates.

CHANGING A VALUE IN A COLUMN TO NULL

There are some special issues involved when dealing with nulls. You already have seen how to add a row in which some of the values are null and how to select rows in which a given column is null. You also must be able to change the value in a column in an existing row to null, as shown in Example 8. Remember that to make this type of change, the affected column must accept nulls. If you specified NOT NULL for the column when you created the table, then changing a value in a column to null is prohibited.

EXAMPLE 8

Change the balance of customer 725 in the LEVEL1_CUSTOMER table to null.

The command for changing a value in a column to null is exactly what it would be for changing any other value. You simply use the value NULL as the replacement value, as shown in Figure 6-15. Notice that the value NULL is *not* enclosed in single quotation marks. If it were, the command would change the balance to the word NULL.

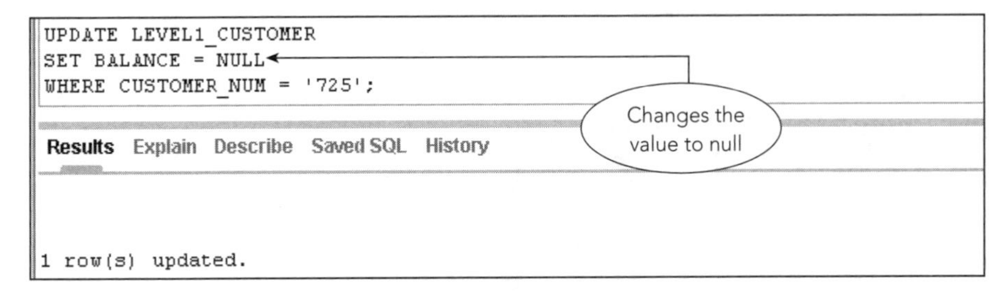

FIGURE 6-15 Changing a value in a column to null

Figure 6-16 shows the data in the LEVEL1_CUSTOMER table after changing the BALANCE column value for customer 725 to null. In Oracle 10*g*, a null value is displayed as a hyphen, as shown in Figure 6-16.

```
SELECT *
FROM LEVEL1_CUSTOMER;
```

Results Explain Describe Saved SQL History

CUSTOMER_NUM	CUSTOMER_NAME	BALANCE	CREDIT_LIMIT	REP_NUM
148	Al's Appliance and Sport	6550	8000	20
356	Ferguson's	5785	7500	65
725	Deerfield's Four Seasons	-	7500	35
842	All Season Sport	8221	7500	20
895	Peter and Margaret's	0	8000	20

5 rows returned in 0.09 seconds CSV Export

Null value

FIGURE 6-16 BALANCE column for customer 725 is null

SQL SERVER USER NOTE

In SQL Server, the word "NULL" appears in the results (without the quotation marks) when a column contains a null value.

CHANGING A TABLE'S STRUCTURE

One of the nicest features of a relational DBMS is the ease with which you can change table structures. In addition to adding new tables to the database and deleting tables that are no longer required, you can add new columns to a table and change the physical characteristics of existing columns. Next, you will see how to accomplish these changes.

You can change a table's structure in SQL by using the **ALTER TABLE** command, as illustrated in the following examples.

EXAMPLE 9

Premiere Products decides to maintain a customer type for each customer in the database. These types are R for regular customers, D for distributors, and S for special customers. Add this information in a new column named CUSTOMER_TYPE in the LEVEL1_CUSTOMER table.

To add a new column, use the **ADD clause** of the ALTER TABLE command. The format for the ALTER TABLE command is the words ALTER TABLE followed by the name of the table to be altered and an appropriate clause. The ADD clause consists of the word ADD followed by the name of the column to be added, followed by the characteristics of the column. Figure 6-17 shows the appropriate ALTER TABLE command for this example.

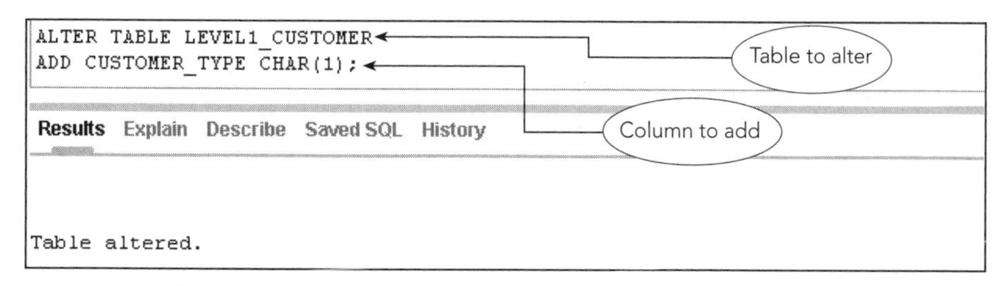

FIGURE 6-17 Adding a column to an existing table

The LEVEL1_CUSTOMER table now contains a column named CUSTOMER_TYPE, a CHAR column with a length of 1. Any new rows added to the table must include values for the new column. Effective immediately, all existing rows also contain this new column. The data in any existing row will contain the new column the next time the row is updated. Any time a row is selected for any reason, however, the system treats the row as though the column is actually present. Thus, to the user, it seems as though the structure was changed immediately.

For existing rows, you must assign some value to the CUSTOMER_TYPE column. The simplest approach (from the point of view of the DBMS, not the user) is to assign the value NULL as a CUSTOMER_TYPE in all existing rows. This process requires the CUSTOMER_TYPE column to accept null values, and some systems actually insist on this. The default for Oracle, Access, and SQL Server is to accept null values.

To change the values in a new column that was added using an ALTER TABLE command, follow the ALTER TABLE command with an UPDATE command like the one shown in Figure 6-18, which sets the CUSTOMER_TYPE value for all rows to R.

```
UPDATE LEVEL1_CUSTOMER
SET CUSTOMER_TYPE = 'R';
```

Omitting the WHERE clause updates all rows

Results Explain Describe Saved SQL History

5 row(s) updated.

FIGURE 6-18 Making the same update for all rows

The SELECT command shown in Figure 6-19 verifies that the value in the CUSTOMER_TYPE column for all rows is R.

```
SELECT *
FROM LEVEL1_CUSTOMER;
```

Results Explain Describe Saved SQL History

CUSTOMER_NUM	CUSTOMER_NAME	BALANCE	CREDIT_LIMIT	REP_NUM	CUSTOMER_TYPE
148	Al's Appliance and Sport	6550	8000	20	R
356	Ferguson's	5785	7500	65	R
725	Deerfield's Four Seasons	-	7500	35	R
842	All Season Sport	8221	7500	20	R
895	Peter and Margaret's	0	8000	20	R

5 rows returned in 0.18 seconds CSV Export

FIGURE 6-19 CUSTOMER_TYPE set to R for all rows

EXAMPLE 10

Two customers in the LEVEL1_CUSTOMER table have a type other than R. Change the types for customers 842 and 148 to S and D, respectively.

Example 9 used an UPDATE command to assign type R to every customer. To change individual types to something other than type R, use the UPDATE command. Figure 6-20 shows the UPDATE command to change customer 842 to customer type S.

```
UPDATE LEVEL1_CUSTOMER
SET CUSTOMER_TYPE = 'S'
WHERE CUSTOMER_NUM = '842';
```

Results Explain Describe Saved SQL History

1 row(s) updated.

FIGURE 6-20 Updating customer 842 to customer type S

Figure 6-21 shows the UPDATE command to change customer 148 to customer type D.

```
UPDATE LEVEL1_CUSTOMER
SET CUSTOMER_TYPE = 'D'
WHERE CUSTOMER_NUM = '148';
```

Results Explain Describe Saved SQL History

1 row(s) updated.

FIGURE 6-21 Updating customer 148 to customer type D

The SELECT command shown in Figure 6-22 shows the results of these UPDATE commands. The customer type for customer 842 is S and the type for customer 148 is D. The type for all other customers is R.

```
SELECT *
FROM LEVEL1_CUSTOMER;
```

Results Explain Describe Saved SQL History

CUSTOMER_NUM	CUSTOMER_NAME	BALANCE	CREDIT_LIMIT	REP_NUM	CUSTOMER_TYPE
148	Al's Appliance and Sport	6550	8000	20	D
356	Ferguson's	5785	7500	65	R
725	Deerfield's Four Seasons	-	7500	35	R
842	All Season Sport	8221	7500	20	S
895	Peter and Margaret's	0	8000	20	R

5 rows returned in 0.00 seconds CSV Export

FIGURE 6-22 Customer types in the LEVEL1_CUSTOMER table after updates

Figure 6-23 uses the DESCRIBE command to display the structure of the LEVEL1_CUSTOMER table, which now includes the CUSTOMER_TYPE column.

```
DESCRIBE LEVEL1_CUSTOMER;
```

DESCRIBE command

New CUSTOMER_TYPE column

Results Explain **Describe** Saved SQL History

Object Type **TABLE** Object **LEVEL1_CUSTOMER**

Table	Column	Data Type	Length	Precision	Scale	Primary Key	Nullable	Default	Comment
LEVEL1_CUSTOMER	CUSTOMER_NUM	Char	3	-	-	1	-	-	-
	CUSTOMER_NAME	Char	35	-	-	-	✓	-	-
	BALANCE	Number	-	8	2	-	✓	-	-
	CREDIT_LIMIT	Number	-	8	2	-	✓	-	-
	REP_NUM	Char	2	-	-	-	✓	-	-
	CUSTOMER_TYPE	Char	1	-	-	-	✓	-	-

1 - 6

FIGURE 6-23 Structure of the LEVEL1_CUSTOMER table

ACCESS USER NOTE

In Access, use the Documenter tool to show the layout of a table.

SQL SERVER USER NOTE

In SQL Server, execute the following command to list all the columns in the LEVEL1_CUSTOMER table:

```
Exec sp_columns LEVEL1_CUSTOMER
```

EXAMPLE 11

The length of the CUSTOMER_NAME column in the LEVEL1_CUSTOMER table is too short. Increase its length to 50 characters. In addition, change the CREDIT_LIMIT column so it cannot accept nulls.

You can change the characteristics of existing columns by using the **MODIFY clause** of the ALTER TABLE command. Figure 6-24 shows the ALTER TABLE command that changes the length of the CUSTOMER_NAME column from 35 to 50 characters.

```
ALTER TABLE LEVEL1_CUSTOMER
MODIFY CUSTOMER_NAME CHAR(50);
```

Results Explain Describe Saved SQL History

```
Table altered.
```

FIGURE 6-24 Changing the length of the CUSTOMER_NAME column in the LEVEL1_CUSTOMER table

SQL SERVER USER NOTE

To change the length of the CUSTOMER_NAME column in SQL Server, use the following ALTER COLUMN clause:

```
ALTER TABLE LEVEL1_CUSTOMER
ALTER COLUMN CUSTOMER_NAME CHAR(50);
```


ACCESS USER NOTE

The version of the ALTER TABLE command shown in Figure 6-24 is not available in Access; to modify the table's structure, make the changes in Design view and save the table.

NOTE

You also can decrease the length of columns, but you might lose some data currently in the column. For example, if you decrease the length of the CUSTOMER_NAME column from 35 to 20 characters, only the first 20 characters of the current customer names will be included. Any characters from position 21 on will be lost. Thus, you should only decrease column lengths when you are positive that you will not lose any data stored in the column.

You can change the length of DECIMAL columns in the same manner that you change the length of CHAR columns.

Figure 6-25 shows the ALTER TABLE command to change the CREDIT_LIMIT column so it does not accept null values.

```
ALTER TABLE LEVEL1_CUSTOMER
MODIFY CREDIT_LIMIT NOT NULL;
```

Results Explain Describe Saved SQL History

```
Table altered.
```

FIGURE 6-25 Changing the CREDIT_LIMIT column in the LEVEL1_CUSTOMER table to reject null values

The DESCRIBE command shown in Figure 6-26 shows the revised structure of the LEVEL1_CUSTOMER table. The length of the CUSTOMER_NAME column is 50 characters.

The dash in the Nullable column for the CREDIT_LIMIT column (instead of a check mark) indicates that the CREDIT_LIMIT column no longer accepts null values.

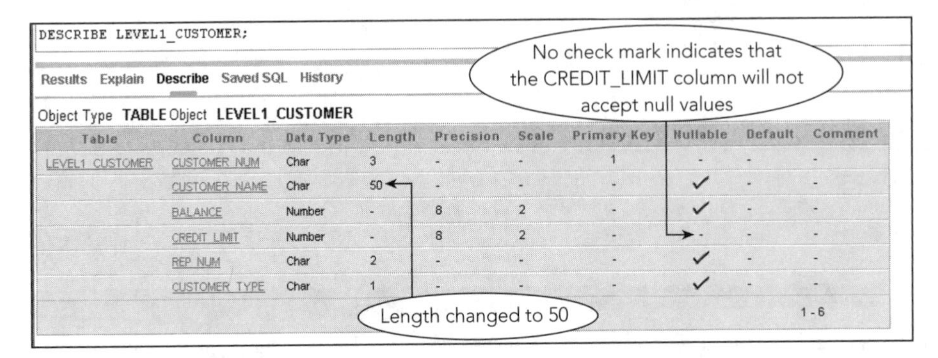

Table	Column	Data Type	Length	Precision	Scale	Primary Key	Nullable	Default	Comment
LEVEL1_CUSTOMER	CUSTOMER_NUM	Char	3	-	-	1	-	-	-
	CUSTOMER_NAME	Char	50	-	-	-	✓	-	-
	BALANCE	Number	-	8	2	-	✓	-	-
	CREDIT_LIMIT	Number	-	8	2	-	-	-	-
	REP_NUM	Char	2	-	-	-	✓	-	-
	CUSTOMER_TYPE	Char	1	-	-	-	✓	-	-

FIGURE 6-26 Revised structure of the LEVEL1_CUSTOMER table

NOTE

You also can use the MODIFY clause of the ALTER TABLE command to change a column that currently rejects null values so that it accepts null values by using NULL in place of NOT NULL in the ALTER TABLE command.

NOTE

If there were existing rows in the LEVEL1_CUSTOMER table in which the CREDIT_LIMIT column was already null, the DBMS would reject the modification to the CREDIT_LIMIT column shown in Figure 6-25 and display an error message indicating that this change is not possible. In this case, you first must use an UPDATE command to change all values that are null to some other value. Then you could alter the table's structure as shown in the figure.

Making Complex Changes

In some cases, you might need to change a table's structure in ways that are either beyond the capabilities of SQL or that are so complex that it would take longer to make the changes than to re-create the table. Perhaps you need to eliminate multiple columns, rearrange the order of several columns, or combine data from two tables into one. For example, if you try to change a column with a data type of VARCHAR to CHAR, SQL still uses VARCHAR when the table contains other variable-length columns. In these situations, you can use a CREATE TABLE command to describe the new table (which must use a different name than the existing table), and then insert values from the existing table into it using the INSERT command combined with an appropriate SELECT command.

DROPPING A TABLE

As you learned in Chapter 3, you can delete a table that is no longer needed by executing the DROP TABLE command.

EXAMPLE 12

Delete the LEVEL1_CUSTOMER table because it is no longer needed in the Premiere Products database.

The command to delete the table is shown in Figure 6-27.

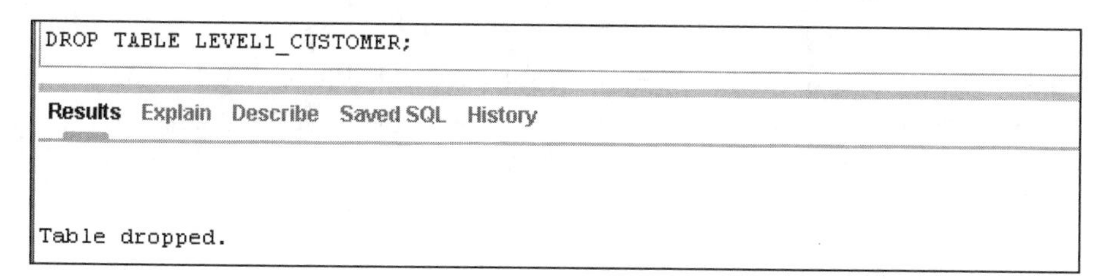

FIGURE 6-27 DROP TABLE command to delete the LEVEL1_CUSTOMER table

When the command shown in Figure 6-27 is executed, the LEVEL1_CUSTOMER table and all its data are permanently removed from the database.

Chapter Summary

- To create a new table from an existing table, first create the new table by using the CREATE TABLE command. Then use an INSERT command containing a SELECT command to select the desired data to be included from the existing table.
- Use the UPDATE command to change existing data in a table.
- Use the INSERT command to add new rows to a table.
- Use the DELETE command to delete existing rows from a table.
- Use the COMMIT command to make updates permanent; use the ROLLBACK command to reverse any updates that have not been committed.
- To change all values in a column to null, use the SET clause followed by the column name, an equal sign, and the word NULL. To change a specific value in a column to null, use a condition to select the row.
- To add a column to a table, use the ALTER TABLE command with an ADD clause.
- To change the characteristics of a column, use the ALTER TABLE command with a MODIFY clause.
- Use the DROP TABLE command to delete a table and all its data.

Key Terms

ADD clause
ALTER TABLE
Autocommit
commit
COMMIT
DELETE
MODIFY clause
roll back
ROLLBACK
transaction
UPDATE

Review Questions

1. Which command creates a new table?
2. Which command and clause adds an individual row to a table?
3. How do you add data from an existing table to another table?
4. Which command changes data in a table?
5. Which command removes rows from a table?
6. In Oracle and in SQL Server, which command makes updates permanent?
7. In Oracle and in SQL Server, which command reverses updates? Which updates are reversed?
8. How do you use the COMMIT and ROLLBACK commands to support transactions?
9. What is the format of the SET clause that changes the value in a column to null in an UPDATE command?
10. Which command and clause adds a column to an existing table?

11. In Oracle and in SQL Server, which command and clause changes the characteristics of an existing column in a table?
12. Which command deletes a table and all its data?
13. Microsoft Access supports make-table queries. What is a make-table query? What SQL statement(s) are equivalent to make-table queries?
14. Use your favorite Web browser and Web search engine to find the SQL command to delete a column in a table. Write the SQL command in Oracle to delete the CUSTOMER_TYPE column from the LEVEL1_CUSTOMER table. Would you use the same command in SQL Server to delete the column? If no, write the command to use in SQL Server.

Exercises

Premiere Products

Use SQL to make the following changes to the Premiere Products database (see Figure 1-2 in Chapter 1). After each change, execute an appropriate query to show that the change was made correctly. If directed to do so by your instructor, use the information provided with the Chapter 3 Exercises to print your output.

1. Create a NONAPPLIANCE table with the structure shown in Figure 6-28.

NONAPPLIANCE

Column	Type	Length	Decimal Places	Nulls Allowed?	Description
PART_NUM	CHAR	4		No	Part number (primary key)
DESCRIPTION	CHAR	15			Part description
ON_HAND	DECIMAL	4	0		Number of units on hand
CLASS	CHAR	2			Item class
PRICE	DECIMAL	6	2		Unit price

FIGURE 6-28 NONAPPLIANCE table layout

2. Insert into the NONAPPLIANCE table the part number, part description, number of units on hand, item class, and unit price from the PART table for each part that is *not* in item class AP.
3. In the NONAPPLIANCE table, change the description of part number AT94 to “Steam Iron.”
4. In the NONAPPLIANCE table, increase the price of each item in item class SG by three percent. (*Hint*: Multiply each price by 1.03.)
5. Add the following part to the NONAPPLIANCE table: part number: TL92; description: Edge Trimmer; number of units on hand: 11; class: HW; and price: 29.95.
6. Delete every part in the NONAPPLIANCE table for which the class is SG.
7. In the NONAPPLIANCE table, change the class for part FD21 to null.
8. Add a column named ON_HAND_VALUE to the NONAPPLIANCE table. The on-hand value is a seven-digit number with two decimal places that represents the product of the number of units on hand and the price. Then set all values of ON_HAND_VALUE to ON_HAND * PRICE.

9. In the NONAPPLIANCE table, increase the length of the DESCRIPTION column to 30 characters.
10. Remove the NONAPPLIANCE table from the Premiere Products database.

Henry Books

Use SQL to make the following changes to the Henry Books database (Figures 1-4 through 1-7 in Chapter 1). After each change, execute an appropriate query to show that the change was made correctly. If directed to do so by your instructor, use the information provided with the Chapter 3 Exercises to print your output.

1. Create a FICTION table with structure shown in Figure 6-29.

FICTION

Column	Type	Length	Decimal Places	Nulls Allowed?	Description
BOOK_CODE	CHAR	4		No	Book code (primary key)
TITLE	CHAR	40			Book title
PUBLISHER_CODE	CHAR	3			Publisher code
PRICE	DECIMAL	4	2		Book price

FIGURE 6-29 FICTION table layout

2. Insert into the FICTION table the book code, book title, publisher code, and price from the BOOK table for only those books having type FIC.
3. The publisher with code LB has decreased the price of its fiction books by four percent. Update the prices in the FICTION table accordingly.
4. Insert a new book into the FICTION table. The book code is 9946, the title is *Cannery Row*, the publisher is PE, and the price is 11.95.
5. Delete the book in the FICTION table having the book code 9883.
6. The price of the book entitled *To Kill a Mockingbird* has been increased to an unknown amount. Change the value in the FICTION table to reflect this change.
7. Add to the FICTION table a new character column named BEST_SELLER that is one character in length. Then set the default value for all columns to N.
8. Change the BEST_SELLER column in the FICTION table to Y for the book entitled *Song of Solomon*.
9. Change the length of the TITLE column in the FICTION table to 50 characters.
10. Change the BEST_SELLER column in the FICTION table to reject nulls.
11. Delete the FICTION table from the database.

Alexamara Marina Group

Use SQL to make the following changes to the Alexamara Marina Group database (Figures 1-8 through 1-12 in Chapter 1). After each change, execute an appropriate query to show that the change was made correctly. If directed to do so by your instructor, use the information provided with the Chapter 3 Exercises to print your output.

1. Create a LARGE_SLIP table with the structure shown in Figure 6-30. (*Hint*: If you have trouble creating the primary key, see Figure 3-31 in Chapter 3.)

LARGE_SLIP

Column	Type	Length	Decimal Places	Nulls Allowed?	Description
MARINA_NUM	CHAR	4		No	Marina number (primary key)
SLIP_NUM	CHAR	4		No	Slip number in the marina (primary key)
RENTAL_FEE	DECIMAL	8	2		Annual rental fee for the slip
BOAT_NAME	CHAR	50			Name of boat currently in the slip
OWNER_NUM	CHAR	4			Number of boat owner renting the slip

FIGURE 6-30 LARGE_SLIP table layout

2. Insert into the LARGE_SLIP table the marina number, slip number, rental fee, boat name, and owner number for those slips whose length is 40 feet.
3. Alexamara has increased the rental fee of each large slip by $150. Update the rental fees in the LARGE_SLIP table accordingly.
4. After increasing the rental fee of each large slip by $150 (Exercise 3), Alexamara decides to decrease the rental fee of any slip whose fee is more than $4,000 by one percent. Update the rental fees in the LARGE_SLIP table accordingly.
5. Insert a new row into the LARGE_SLIP table. The marina number is 1, the slip number is A4, the rental fee is $3,900, the boat name is *Bilmore*, and the owner number is FE82.
6. Delete all slips in the LARGE_SLIP table for which the owner number is TR72.
7. The name of the boat in marina 1 and slip A1 is in the process of being changed to an unknown name. Change the name of this boat in the LARGE_SLIP table to null.
8. Add to the LARGE_SLIP table a new character column named CHARTER that is one character in length. (This column will indicate whether the boat is available for chartering.) Set the value for the CHARTER column on all rows to N.
9. Change the CHARTER column in the LARGE_SLIP table to Y for the slip containing the boat named *Our Toy*.
10. Change the length of the BOAT_NAME column in the LARGE_SLIP table to 60 characters.
11. Change the RENTAL_FEE column in the LARGE_SLIP table to reject nulls.
12. Delete the LARGE_SLIP table from the database.

CHAPTER 7

DATABASE ADMINISTRATION

LEARNING OBJECTIVES

Objectives

- Understand, create, and drop views
- Recognize the benefits of using views
- Use a view to update data
- Grant and revoke users' database privileges
- Understand the purpose, advantages, and disadvantages of using an index
- Create, use, and drop an index
- Understand and obtain information from the system catalog
- Use integrity constraints to control data entry

INTRODUCTION

There are some special issues involved in managing a database. This process, often called **database administration**, is especially important when more than one person uses the database. In a business organization, a person or an entire group known as the **database administrator** is charged with managing the database.

In Chapter 6, you learned about one function of the database administrator: changing the structure of a database. In this chapter, you will see how the database administrator can give each user his or her own view of the database. You will use the GRANT and REVOKE commands to assign different database privileges to different users. You will use indexes to improve database performance. You will learn how a

DBMS stores information about the database structure in an object called the system catalog and how to access that information. Finally, you will learn how to specify integrity constraints that establish rules for the data in the database.

CREATING AND USING VIEWS

Most DBMSs support the creation of views. A **view** is a program's or an individual user's picture of the database. The existing, permanent tables in a relational database are called **base tables**. A view is a derived table because the data in it comes from one or more base tables. To the user, a view appears to be an actual table, but it is not. In many cases, a user can examine table data using a view. Because a view usually includes less information than the full database, its use can represent a great simplification. Views also provide a measure of security, because omitting sensitive tables or columns from a view renders them unavailable to anyone accessing the database through the view.

To help you understand the concept of a view, suppose that Juan is interested in the part number, part description, units on hand, and unit price of parts in item class HW. He is not interested in any other columns in the PART table, nor is he interested in any rows that correspond to parts in other item classes. Viewing this data would be simpler for Juan if the other rows and columns were not even present. Although you cannot change the structure of the PART table and omit some of its rows just for Juan, you can do the next best thing. You can provide him with a view that consists of only the rows and columns that he needs to access.

A view is defined by creating a **defining query**, which indicates the rows and columns to include in the view. The SQL command (or the defining query) to create the view for Juan is illustrated in Example 1.

EXAMPLE 1

Create a view named HOUSEWARES that consists of the part number, part description, units on hand, and unit price of each part in item class HW.

To create a view, use the **CREATE VIEW** command, which includes the words CREATE VIEW, followed by the name of the view, the word AS, and then a query. The CREATE VIEW command shown in Figure 7-1 creates a view of the PART table that contains only the specified columns.

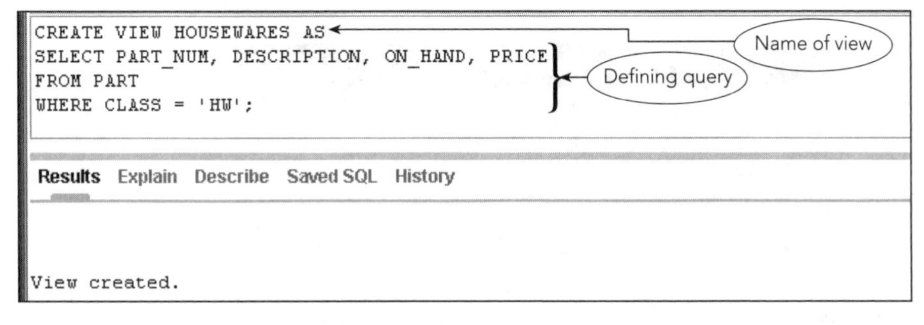

FIGURE 7-1 Creating the HOUSEWARES view

ACCESS USER NOTE

Access does not support the CREATE VIEW command. To create a view in Access, create a query to define the view, and then save the query object in the database using the view's name (for example, HOUSEWARES).

Given the current data in the Premiere Products database, the HOUSEWARES view contains the data shown in Figure 7-2.

HOUSEWARES

PART_NUM	DESCRIPTION	ON_HAND	PRICE
AT94	Iron	50	$24.95
DL71	Cordless Drill	21	$129.95
FD21	Stand Mixer	22	$159.95

FIGURE 7-2 HOUSEWARES view

The data does not actually exist in this form, nor will it *ever* exist in this form. It is tempting to think that when Juan uses this view, the query is executed and produces some sort of temporary table, named HOUSEWARES, that Juan can access, but this is *not* what actually happens. Instead, the query acts as a sort of "window" into the database, as shown in Figure 7-3. As far as Juan is concerned, the entire database is just the darker shaded portion of the PART table. Juan can see any change that affects the darker portion of the PART table, but he is totally unaware of any other changes that are made in the database.

PART

PART_NUM	DESCRIPTION	ON_HAND	CLASS	WAREHOUSE	PRICE
AT94	Iron	50	HW	3	$24.95
BV06	Home Gym	45	SG	2	$794.95
CD52	Microwave Oven	32	AP	1	$165.00
DL71	Cordless Drill	21	HW	3	$129.95
DR93	Gas Range	8	AP	2	$495.00
DW11	Washer	12	AP	3	$399.99
FD21	Stand Mixer	22	HW	3	$159.95
KL62	Dryer	12	AP	1	$349.95
KT03	Dishwasher	8	AP	3	$595.00
KV29	Treadmill	9	SG	2	$1,390.00

FIGURE 7-3 Juan's view of the PART table

When you create a query that involves a view, the DBMS changes the query to one that selects data from the table(s) in the database that created the view. For example, suppose Juan creates the query shown in Figure 7-4.

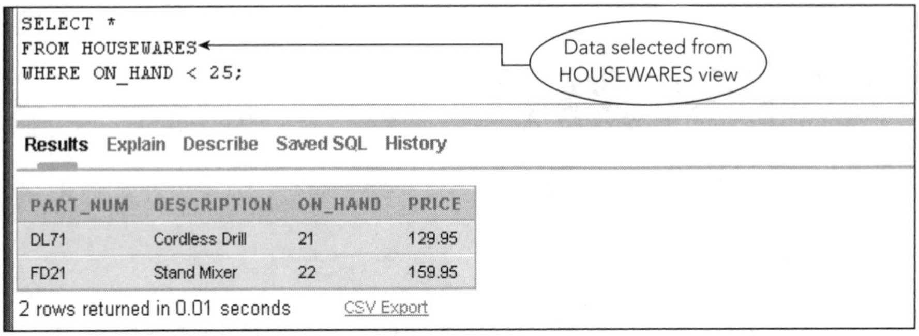

```
SELECT *
FROM HOUSEWARES
WHERE ON_HAND < 25;
```


PART_NUM	DESCRIPTION	ON_HAND	PRICE
DL71	Cordless Drill	21	129.95
FD21	Stand Mixer	22	159.95

2 rows returned in 0.01 seconds CSV Export

FIGURE 7-4 Using the HOUSEWARES view

The DBMS does not execute the query in this form. Instead, it merges the query Juan entered with the query that creates the view to form the query that is actually executed. When the DBMS merges the query that creates the view with Juan's query to select rows for which the ON_HAND value is less than 25, the query that the DBMS actually executes is:

```
SELECT PART_NUM, DESCRIPTION, ON_HAND, PRICE
FROM PART
WHERE CLASS = 'HW'
AND ON_HAND < 25;
```

In the query that the DBMS executes, the FROM clause lists the PART table rather than the HOUSEWARES view, the SELECT clause lists columns from the PART table instead of * to select all columns from the HOUSEWARES view, and the WHERE clause contains a compound condition to select only those parts in the HW class (as Juan sees in the

HOUSEWARES view) and only those parts with ON_HAND values of less than 25. This new query is the one that the DBMS actually executes.

Juan, however, is unaware that this activity is taking place. To Juan, it seems that he is really using a table named HOUSEWARES. One advantage of this approach is that because the HOUSEWARES view never exists in its own right, any update to the PART table is *immediately* available in the HOUSEWARES view. If the HOUSEWARES view were really a table, this immediate update would not be possible.

You also can assign column names that are different from those in the base table, as illustrated in the next example.

EXAMPLE 2

Create a view named HSEWRES that consists of the part number, part description, units on hand, and unit price of all parts in item class HW. In this view, change the names of the PART_NUM, DESCRIPTION, ON_HAND, and PRICE columns to PNUM, DSC, OH, and PRCE, respectively.

When renaming columns, you include the new column names in parentheses following the name of the view, as shown in Figure 7-5. In this case, anyone accessing the HSEWRES view will refer to PART_NUM as PNUM, to DESCRIPTION as DSC, to ON_HAND as OH, and to PRICE as PRCE.

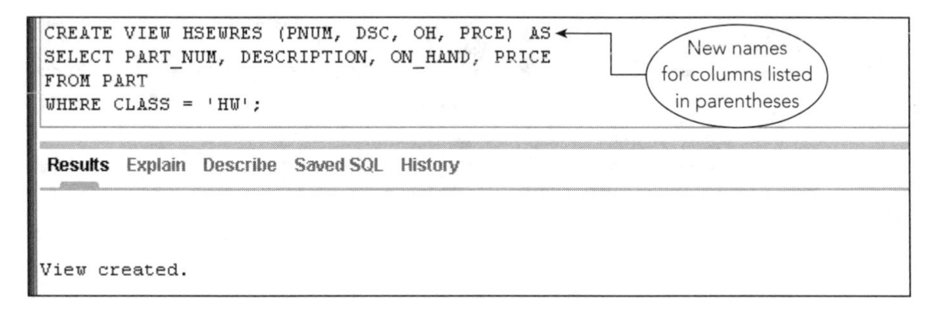

FIGURE 7-5 Renaming columns when creating a view

If you select all columns from the HSEWRES view, the output displays the new column names, as shown in Figure 7-6.

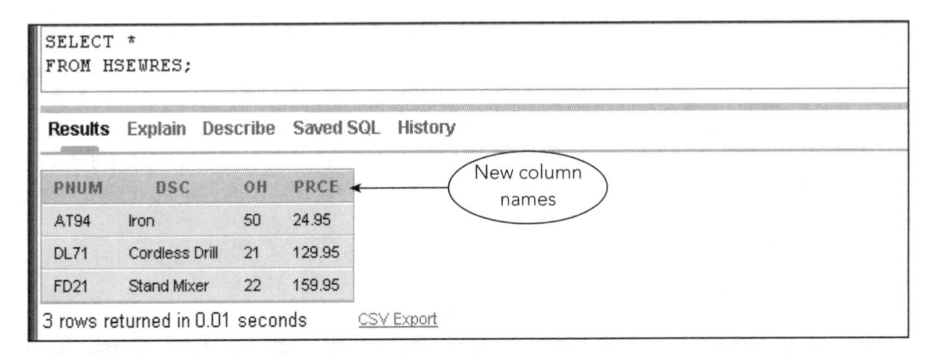
```
SELECT *
FROM HSEWRES;
```

Results Explain Describe Saved SQL History

PNUM	DSC	OH	PRCE
AT94	Iron	50	24.95
DL71	Cordless Drill	21	129.95
FD21	Stand Mixer	22	159.95

3 rows returned in 0.01 seconds CSV Export

FIGURE 7-6 Data in the HSEWRES view

ACCESS USER NOTE

To change column names in Access, use AS clauses in the query design (for example, SELECT PART_NUM AS PNUM, DESCRIPTION AS DSC, and so on).

The HSEWRES view is an example of a **row-and-column subset view** because it consists of a subset of the rows and columns in some base table—in this case, in the PART table. Because the defining query can be any valid SQL query, a view also can join two or more tables or involve statistics. The next example illustrates a view that joins two tables.

EXAMPLE 3

Create a view named REP_CUST consisting of the sales rep number (named RNUM), sales rep last name (named RLAST), sales rep first name (named RFIRST), customer number (named CNUM), and customer name (named CNAME) for all sales reps and matching customers in the REP and CUSTOMER tables.

The command to create this view appears in Figure 7-7.

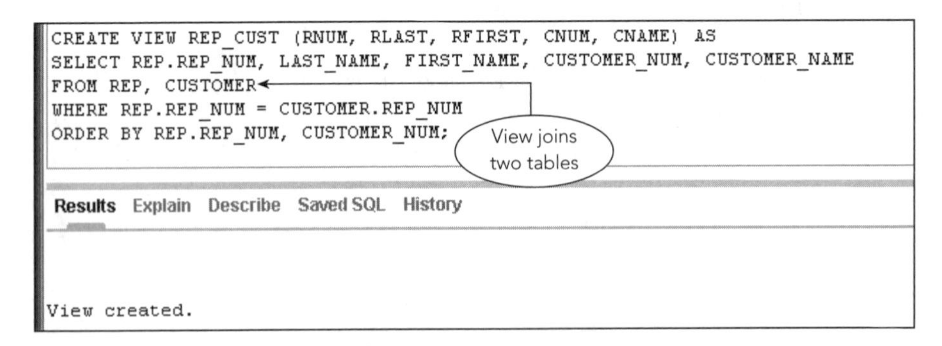
```
CREATE VIEW REP_CUST (RNUM, RLAST, RFIRST, CNUM, CNAME) AS
SELECT REP.REP_NUM, LAST_NAME, FIRST_NAME, CUSTOMER_NUM, CUSTOMER_NAME
FROM REP, CUSTOMER
WHERE REP.REP_NUM = CUSTOMER.REP_NUM
ORDER BY REP.REP_NUM, CUSTOMER_NUM;
```

Results Explain Describe Saved SQL History

View created.

FIGURE 7-7 Creating the REP_CUST view

Given the current data in the Premiere Products database, the REP_CUST view contains the data shown in Figure 7-8.

```
SELECT *
FROM REP_CUST;
```

Results Explain Describe Saved SQL History

RNUM	RLAST	RFIRST	CNUM	CNAME
20	Kaiser	Valerie	148	Al's Appliance and Sport
20	Kaiser	Valerie	524	Kline's
20	Kaiser	Valerie	842	All Season
35	Hull	Richard	282	Brookings Direct
35	Hull	Richard	408	The Everything Shop
35	Hull	Richard	687	Lee's Sport and Appliance
35	Hull	Richard	725	Deerfield's Four Seasons
65	Perez	Juan	356	Ferguson's
65	Perez	Juan	462	Bargains Galore
65	Perez	Juan	608	Johnson's Department Store

10 rows returned in 0.02 seconds CSV Export

FIGURE 7-8 Data in the REP_CUST view

SQL SERVER USER NOTE

SQL Server does not support the ORDER BY clause in a CREATE VIEW command. When you need to order the query results, insert an ORDER BY clause in the SELECT command when you query the view. For example, the following SELECT command retrieves all records in the REP_CUST view ordered by rep number and customer number:

```
SELECT *
FROM REP_CUST
ORDER BY RNUM, CNUM
```

A view also can involve statistics, as illustrated in Example 4.

EXAMPLE 4

Create a view named CRED_CUST that consists of each credit limit (CREDIT_LIMIT) and the number of customers having this credit limit (NUM_CUSTOMERS). Sort the credit limits in ascending order.

The command shown in Figure 7-9 creates this view.

```
CREATE VIEW CRED_CUST (CREDIT_LIMIT, NUM_CUSTOMERS) AS
SELECT CREDIT_LIMIT, COUNT(*)
FROM CUSTOMER
GROUP BY CREDIT_LIMIT
ORDER BY CREDIT_LIMIT;
```

Results Explain Describe Saved SQL History

View created.

FIGURE 7-9 Creating the CRED_CUST view

The SELECT command shown in Figure 7-10 displays the current data in the Premiere Products database for this view.

```
SELECT *
FROM CRED_CUST;
```

Results Explain Describe Saved SQL History

CREDIT_LIMIT	NUM_CUSTOMERS
5000	2
7500	4
10000	3
15000	1

4 rows returned in 0.01 seconds CSV Export

FIGURE 7-10 Data in the CRED_CUST view

The use of views provides several benefits. First, views provide data independence. When the database structure changes (by adding columns or changing the way objects are related, for example) in such a way that the view still can be derived from existing data, the user can access and use the same view. If adding extra columns to tables in the database is the only change, and these columns are not required by the view's user, the defining query might not even need to be changed for the user to continue using the view. If table relationships are changed, the defining query might be different, but because users are not aware of the defining query, they are unaware of this difference. Users continue accessing the database through the same view, as though nothing has changed. For example, suppose customers are assigned to territories, each territory is assigned to a single sales rep, a sales rep can have more than one territory, and a customer is represented by the sales rep who covers the customer's assigned territory. To implement these changes, you might choose to restructure the database as follows:

```
REP(REP_NUM, LAST_NAME, FIRST_NAME, STREET, CITY,
     STATE, ZIP, COMMISSION, RATE)
TERRITORY(TERRITORY_NUM, DESCRIPTION, REP_NUM)
CUSTOMER(CUSTOMER_NUM, CUSTOMER_NAME, STREET, CITY,
     STATE, ZIP, BALANCE, CREDIT_LIMIT, TERRITORY_NUM)
```

Assuming that the REP_CUST view created in Figure 7-7 is still required, you could change the defining query as follows:


```
CREATE VIEW REP_CUST (RNUM, RLAST, RFIRST,
     CNUM, CNAME) AS
SELECT REP.REP_NUM, REP.LAST_NAME, REP.FIRST_NAME,
     CUSTOMER_NUM, CUSTOMER_NAME
FROM REP, TERRITORY, CUSTOMER
WHERE REP.REP_NUM = TERRITORY.REP_NUM
AND TERRITORY.TERRITORY_NUM = CUSTOMER.TERRITORY_NUM;
```

This view's user still can retrieve the number and name of a sales rep together with the number and name of each customer the sales rep represents. The user is unaware, however, of the new structure in the database.

The second benefit of using views is that different users can see the same data in different ways through their own views. In other words, you can customize the display of data to meet each user's needs.

The final benefit of using views is that a view can contain only those columns required by a given user. This practice has two advantages. First, because the view usually contains fewer columns than the overall database and is conceptually a single table, rather than a collection of tables, a view greatly simplifies the user's perception of the database. Second, views provide a measure of security. Columns that are not included in the view are not accessible to the view's user. For example, omitting the BALANCE column from a view ensures that the view's user cannot access any customer's balance. Likewise, rows that are not included in the view are not accessible. A user of the HOUSEWARES view, for example, cannot obtain any information about parts in the AP or SG classes.

USING A VIEW TO UPDATE DATA

The benefits of using views hold true only when views are used for retrieval purposes. When updating the database, the issues involved in updating data through a view depend on the type of view, as you will see next.

Updating Row-and-Column Subset Views

Consider the row-and-column subset view for the HOUSEWARES view. There are columns in the underlying base table (PART) that are not present in the view. If you attempt to add a row with the data ('BB99','PAN',50,14.95), the DBMS must determine how to enter the data in those columns from the PART table that are not included in the HOUSEWARES view (CLASS and WAREHOUSE). In this case, it is clear what data to enter in the CLASS column. According to the view definition, all rows are item class HW, but it is not clear what data to enter in the WAREHOUSE column. The only possibility would be NULL. Therefore, if every column not included in a view can accept nulls, you can add new rows using the INSERT command. There is another problem, however. Suppose the user attempts to add a row to the HOUSEWARES view containing the data ('BV06','Waffle Maker',5,29.95). Because part number BV06 already exists in the PART table, the system *must* reject this attempt. Because this part is not in item class HW (and therefore is not in the HOUSEWARES view), this rejection certainly will seem strange to the user, because there is no such part in the user's view.

On the other hand, updates or deletions cause no particular problem in this view. If the description of part number FD21 changes from Stand Mixer to Pan, this change is made in the PART table. If part number DL71 is deleted, this deletion occurs in the PART table. One surprising change could take place, however. Suppose that the CLASS column is included in the HOUSEWARES view and a user changes the class of part number AT94 from HW to AP. Because this item would no longer satisfy the criterion for being included in the HOUSEWARES view, part number AT94 would disappear from the user's view!

Although there are problems to overcome when updating row-and-column subset views, it seems possible to update the database through the HOUSEWARES view. This does not mean that *any* row-and-column subset view is updatable, however. Consider the REP_CRED view shown in Figure 7-11. (The DISTINCT operator is used to omit duplicate rows from the view.)

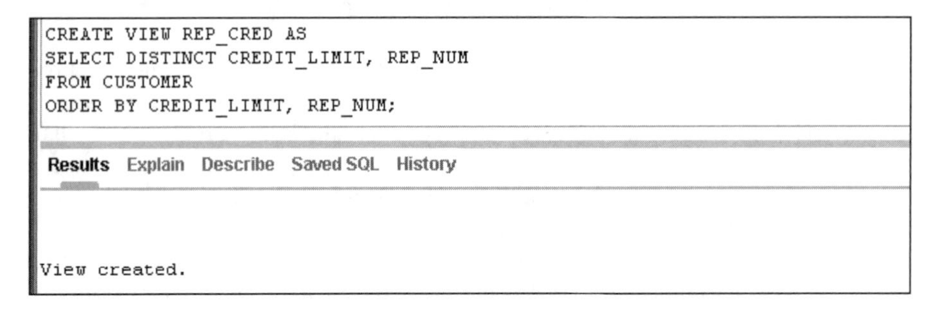

FIGURE 7-11 Creating the REP_CRED view

Figure 7-12 shows the data in the REP_CRED view.

```
SELECT *
FROM REP_CRED;
```

Results Explain Describe Saved SQL History

CREDIT_LIMIT	REP_NUM
5000	35
7500	20
7500	35
7500	65
10000	35
10000	65
15000	20

7 rows returned in 0.01 seconds CSV Export

FIGURE 7-12 Data in the REP_CRED view

How would you add the row 15000,'35' to this view? In the underlying base table (CUSTOMER), at least one customer must be added whose credit limit is $15,000 and whose sales rep number is 35, but which customer is it? You cannot leave the other columns null in this case, because one of them is CUSTOMER_NUM, which is the base table's primary key. What would it mean to change the row 5000,'35' to 15000,'35'? Would it mean changing the credit limit to $15,000 for each customer represented by sales rep number 35 that currently has a credit limit of $5,000? Would it mean changing the credit limit of one of these customers and deleting the rest? What would it mean to delete the row 5000,'35'? Would it mean deleting all customers with credit limits of $5,000 and represented by sales rep number 35, or would it mean assigning these customers a different sales rep or a different credit limit?

Why does the REP_CRED view involve a number of serious problems that are not present in the HOUSEWARES view? The basic reason is that the HOUSEWARES view includes, as one of its columns, the primary key of the underlying base table, but the REP_CRED view does not. A row-and-column subset view that contains the primary key of the underlying base table is updatable (subject, of course, to some of the concerns already discussed).

Updating Views Involving Joins

In general, views that involve joins of base tables can cause problems when updating data. Consider the relatively simple REP_CUST view, for example, described earlier (see Figures 7-7 and 7-8). The fact that some columns in the underlying base tables are not included in this view presents some of the same problems discussed earlier. Assuming that you can overcome these problems by using nulls, there are more serious problems when attempting to update the database through this view. On the surface, changing the row ('35','Hull','Richard', '282','Brookings Direct') to ('35','Baldwin','Sara','282','Brookings Direct'), might not appear to pose any problems other than some inconsistency in the data. (In the new version of the row, the name of sales rep 35 is Sara Baldwin; whereas in the fourth row in the table, the name of sales rep 35, *the same sales rep*, is Richard Hull.)

The problem is actually more serious than that—making this change is not possible. The name of a sales rep is stored only once in the underlying REP table. Changing the name of sales rep 35 from Richard Hull to Sara Baldwin in this one row of the view causes the change to be made to the single row for sales rep 35 in the REP table. Because the view simply displays data from the base tables, for each row on which the sales rep number is 35, the sales rep name is now Sara Baldwin. In other words, it appears that the same change has been made in the other rows. In this case, this change ensures consistency in the data. In general, however, the unexpected changes caused by an update are not desirable.

Before concluding the topic of views that involve joins, you should note that all joins do not create the preceding problem. When two base tables have the same primary key and the primary key is used as the join column, updating the database using the view is not a problem. For example, suppose the actual database contains two tables (REP_DEMO and REP_FIN) instead of one table (REP). Figure 7-13 shows the data in the REP_DEMO table.

```
SELECT *
FROM REP_DEMO;
```

Results Explain Describe Saved SQL History

REP_NUM	LAST_NAME	FIRST_NAME	STREET	CITY	STATE	ZIP
20	Kaiser	Valerie	624 Randall	Grove	FL	33321
35	Hull	Richard	532 Jackson	Sheldon	FL	33553
65	Perez	Juan	1626 Taylor	Fillmore	FL	33336

3 rows returned in 0.00 seconds CSV Export

FIGURE 7-13 Data in the REP_DEMO table

Figure 7-14 shows the data in the REP_FIN table.

```
SELECT *
FROM REP_FIN;
```

Results Explain Describe Saved SQL History

REP_NUM	COMMISSION	RATE
20	20542.5	.05
35	39216	.07
65	23487	.05

3 rows returned in 0.02 seconds CSV Export

FIGURE 7-14 Data in the REP_FIN table

What was once a single table in the original Premiere Products design has been divided into two separate tables. Users who need to see the rep data in a single table can use a view named SALES_REP that joins these two tables using the REP_NUM column. The defining query for the SALES_REP view appears in Figure 7-15.

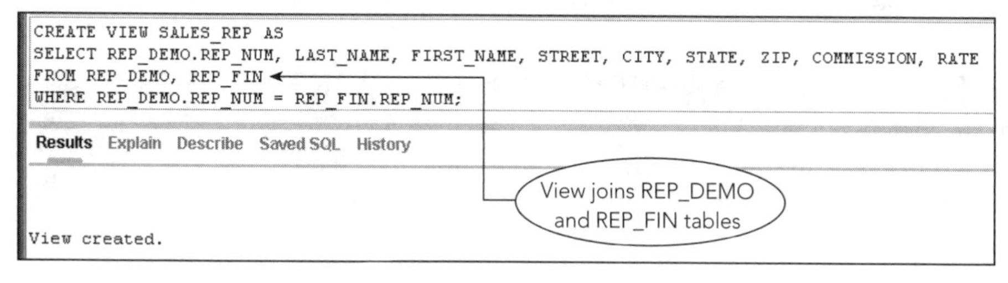

```
CREATE VIEW SALES_REP AS
SELECT REP_DEMO.REP_NUM, LAST_NAME, FIRST_NAME, STREET, CITY, STATE, ZIP, COMMISSION, RATE
FROM REP_DEMO, REP_FIN
WHERE REP_DEMO.REP_NUM = REP_FIN.REP_NUM;
```


FIGURE 7-15 Creating the SALES_REP view

The data in the SALES_REP view appears in Figure 7-16.

```
SELECT *
FROM SALES_REP;
```

Results Explain Describe Saved SQL History

REP_NUM	LAST_NAME	FIRST_NAME	STREET	CITY	STATE	ZIP	COMMISSION	RATE
20	Kaiser	Valerie	624 Randall	Grove	FL	33321	20542.5	.05
35	Hull	Richard	532 Jackson	Sheldon	FL	33553	39216	.07
65	Perez	Juan	1626 Taylor	Fillmore	FL	33336	23487	.05

3 rows returned in 0.02 seconds CSV Export

FIGURE 7-16 Data in the SALES_REP view

It is easy to update the SALES_REP view. To add a row, use an INSERT command to add a row to each underlying base table. To update data in a row, make the change in the appropriate base table. To delete a row from the view, delete the corresponding rows from both underlying base tables.

Q & A

Question: How would you add the row ('10','Peters','Jean','14 Brink','Holt','FL','46223', 107.50,0.05) to the SALES_REP view?
Answer: Use an INSERT command to add the row ('10','Peters','Jean','14 Brink','Holt','FL','46223') to the REP_DEMO table, and then use another INSERT command to add the row ('10',107.50,0.05) to the REP_FIN table.

Q & A

Question: How would you change the name of sales rep 20 to Valerie Lewis?
Answer: Use an UPDATE command to change the name in the REP_DEMO table.

Q & A

Question: How would you change Valerie's commission rate to 0.06?
Answer: Use an UPDATE command to change the rate in the REP_FIN table.

Q & A

Question: How would you delete sales rep 35 from the REP table?
Answer: Use a DELETE command to delete sales rep 35 from *both* the REP_DEMO *and* REP_FIN tables.

Updates (additions, changes, or deletions) to the SALES_REP view do not cause any problems. The main reason that the SALES_REP view is updatable—and that other views involving joins might not be updatable—is that this view is derived from joining two base tables *on the primary key of each table*. In contrast, the REP_CUST view is created by joining two tables by matching the primary key of one table with a column that is *not* the primary key in the other table. When neither of the join columns in a view is a primary key column, users will encounter even more severe problems when attempting to make updates.

Updating Views Involving Statistics

A view that involves statistics calculated from one or more base tables is the most troublesome view when attempting to update data. Consider the CRED_CUST view, for example (see Figure 7-10). How would you add the row 9000,3 to indicate that there are three customers that have credit limits of $9,000 each? Likewise, changing the row 5000,2 to 5000,5 means you are adding three new customers with credit limits of $5,000 each, for a total of five customers. Clearly these are impossible tasks; you cannot add rows to a view that includes calculations.

DROPPING A VIEW

When a view is no longer needed, you can remove it using the **DROP VIEW** command.

EXAMPLE 5

The HSEWRES view is no longer necessary, so delete it.

The command to delete a view is DROP VIEW as shown in Figure 7-17.

```
DROP VIEW HSEWRES;
```

Results Explain Describe Saved SQL History

```
View dropped.
```

FIGURE 7-17 Dropping a view

ACCESS USER NOTE

Access does not support the DROP VIEW command. To drop a view, delete the query object you saved when you created the view.

SECURITY

Security is the prevention of unauthorized access to a database. Within an organization, the database administrator determines the types of access various users need to the database. Some users might need to retrieve and update anything in the database. Other users might need to retrieve any data from the database but not make any changes to it. Still other users might need to access only a portion of the database. For example, Bill might need to retrieve and update customer data, but does not need to access data about sales reps, orders, order lines, or parts. Valerie might need to retrieve part data and nothing else. Sam might need to retrieve and update data on parts in the HW class, but does not need to retrieve data in any other classes.

After the database administrator has determined the access different users of the database need, the DBMS enforces these access rules by whatever security mechanism the DBMS supports. You can use SQL to enforce two security mechanisms. You already have seen that views furnish a certain amount of security; when users are accessing the database through a view, they cannot access any data that is not included in the view. The main mechanism for providing access to a database, however, is the **GRANT** command.

The basic idea of the GRANT command is that the database administrator can grant different types of privileges to users and then revoke them later, if necessary. These privileges include the right to select, insert, update, and delete table data. You can grant and revoke user privileges using the GRANT and REVOKE commands. The following examples illustrate various uses of the GRANT command when the named users already exist in the database.

NOTE

Do not execute the commands in this section unless your instructor asks you to do so.

EXAMPLE 6

User Johnson must be able to retrieve data from the REP table.

The following GRANT command permits a user named Johnson to execute SELECT commands for the REP table:

```
GRANT SELECT ON REP TO JOHNSON;
```

EXAMPLE 7

Users Smith and Brown must be able to add new parts to the PART table.

The following GRANT command permits two users named Smith and Brown to execute INSERT commands for the PART table. Notice that a comma separates the user names:

```
GRANT INSERT ON PART TO SMITH, BROWN;
```

EXAMPLE 8

User Anderson must be able to change the name and street address of customers.

The following GRANT command permits a user named Anderson to execute UPDATE commands involving the CUSTOMER_NAME and STREET columns in the CUSTOMER table. Notice that the SQL command includes the column names in parentheses before the ON clause:

```
GRANT UPDATE (CUSTOMER_NAME, STREET) ON CUSTOMER TO
ANDERSON;
```

EXAMPLE 9

User Thompson must be able to delete order lines.

The following GRANT command permits a user named Thompson to execute DELETE commands for the ORDER_LINE table:

```
GRANT DELETE ON ORDER_LINE TO THOMPSON;
```

EXAMPLE 10

Every user must be able to retrieve part numbers, part descriptions, and item classes.

The GRANT command to indicate that all users can retrieve data using a SELECT command includes the word PUBLIC, as follows:

```
GRANT SELECT (PART_NUM, DESCRIPTION, CLASS) ON PART TO
PUBLIC;
```


EXAMPLE 11

User Roberts must be able to create an index on the REP table.

You will learn about indexes and their uses in the next section. The following GRANT command permits a user named Roberts to create an index on the REP table:

```
GRANT INDEX ON REP TO ROBERTS;
```

EXAMPLE 12

User Thomas must be able to change the structure of the CUSTOMER table.

The following GRANT command permits a user named Thomas to execute ALTER commands for the CUSTOMER table so he can change the table's structure:

```
GRANT ALTER ON CUSTOMER TO THOMAS;
```

EXAMPLE 13

User Wilson must have all privileges for the REP table.

The GRANT command to indicate that a user has all privileges includes the ALL privilege, as follows:

```
GRANT ALL ON REP TO WILSON;
```

The privileges that a database administrator can grant are SELECT to retrieve data, UPDATE to change data, DELETE to delete data, INSERT to add new data, INDEX to create an index, and ALTER to change the table structure. The database administrator usually assigns privileges. Normally, when the database administrator grants a particular privilege to a user, the user cannot pass that privilege along to other users. When the user needs to be able to pass the privilege to other users, the GRANT command must include the

WITH GRANT OPTION clause. This clause grants the indicated privilege to the user and also permits the user to grant the same privileges (or a subset of them) to other users.

The database administrator uses the **REVOKE** command to revoke privileges from users. The format of the REVOKE command is essentially the same as that of the GRANT command, but with two differences: the word GRANT is replaced by the word REVOKE, and the word TO is replaced by the word FROM. In addition, the clause WITH GRANT OPTION obviously is not meaningful as part of a REVOKE command. Incidentally, the revoke cascades, so if Johnson is granted privileges WITH GRANT OPTION and then Johnson grants these same privileges to Smith, revoking the privileges from Johnson revokes Smith's privileges at the same time. Example 14 illustrates the use of the REVOKE command.

EXAMPLE 14

User Johnson is no longer allowed to retrieve data from the REP table.

The following REVOKE command revokes the SELECT privilege for the REP table from the user named Johnson:

```
REVOKE SELECT ON REP FROM JOHNSON;
```

The database administrator can also apply the GRANT and REVOKE commands to views to restrict access to only certain rows within tables.

INDEXES

When you query a database, you are usually searching for a row (or collection of rows) that satisfies some condition. Examining every row in a table to find the ones you need often takes too much time to be practical, especially in tables with thousands of rows. Fortunately, you can create and use an index to speed up the searching process significantly. An index in SQL is similar to an index in a book. When you want to find a discussion of a given topic in a book, you could scan the entire book from start to finish, looking for references to the topic you need. More than likely, however, you would not have to resort to this technique. If the book has a good index, you could use it to identify the pages on which your topic is discussed.

In a DBMS, the main mechanism for increasing the efficiency with which data is retrieved from the database is the **index**. Conceptually, these indexes are very much like the index in a book. Consider Figure 7-18, for example, which shows the CUSTOMER table for Premiere Products together with one extra column named ROW_NUMBER. This extra column gives the location of the row in the table (customer 148 is the first row in the table and is on row 1, customer 282 is on row 2, and so on). The DBMS—not the user—automatically assigns and uses these row numbers, and that is why you do not see them.

CUSTOMER

ROW_NUMBER	CUSTOMER_NUM	CUSTOMER_NAME	STREET	CITY	STATE	ZIP	BALANCE	CREDIT_LIMIT	REP_NUM
1	148	Al's Appliance and Sport	2837 Greenway	Fillmore	FL	33336	$6,550.00	$7,500.00	20
2	282	Brookings Direct	3827 Devon	Grove	FL	33321	$431.50	$10,000.00	35
3	356	Ferguson's	382 Wildwood	Northfield	FL	33146	$5,785.00	$7,500.00	65
4	408	The Everything Shop	1828 Raven	Crystal	FL	33503	$5,285.25	$5,000.00	35
5	462	Bargains Galore	3829 Central	Grove	FL	33321	$3,412.00	$10,000.00	65
6	524	Kline's	838 Ridgeland	Fillmore	FL	33336	$12,762.00	$15,000.00	20
7	608	Johnson's Department Store	372 Oxford	Sheldon	FL	33553	$2,106.00	$10,000.00	65
8	687	Lee's Sport and Appliance	282 Evergreen	Altonville	FL	32543	$2,851.00	$5,000.00	35
9	725	Deerfield's Four Seasons	282 Columbia	Sheldon	FL	33553	$248.00	$7,500.00	35
10	842	All Season	28 Lakeview	Grove	FL	33321	$8,221.00	$7,500.00	20

FIGURE 7-18 CUSTOMER table with row numbers

To access a customer's row using its customer number, you might create and use an index, as shown in Figure 7-19. The index has two columns: the first column contains a customer number, and the second column contains the number of the row on which the customer number is found. To find a customer, look up the customer's number in the first column in the index. The value in the second column indicates which row to retrieve from the CUSTOMER table, then the row for the desired customer is retrieved.

CUSTOMER_NUM Index

CUSTOMER_NUM	ROW_NUMBER
148	1
282	2
356	3
408	4
462	5
524	6
608	7
687	8
725	9
842	10

FIGURE 7-19 Index for the CUSTOMER table on the CUSTOMER_NUM column

Because customer numbers are unique, there is only a single row number in this index. This is not always the case, however. Suppose you need to access all customers with a specific credit limit or all customers represented by a specific sales rep. You might choose to create and use an index on the CREDIT_LIMIT column and an index on the REP_NUM column, as shown in Figure 7-20. In the CREDIT_LIMIT index, the first column contains a credit limit and the second column contains the numbers of *all* rows on which that credit limit appears. The REP_NUM index is similar, except that the first column contains a sales rep number.

CREDIT_LIMIT Index

CREDIT_LIMIT	ROW_NUMBER
$5,000.00	4, 8
$7,500.00	1, 3, 9, 10
$10,000.00	2, 5, 7
$15,000.00	6

REP_NUM Index

REP_NUM	ROW_NUMBER
20	1, 6, 10
35	2, 4, 8, 9
65	3, 5, 7

FIGURE 7-20 Indexes for the CUSTOMER table on the CREDIT_LIMIT and REP_NUM columns

Q & A

Question: How would you use the index shown in Figure 7-20 to find every customer with a $10,000 credit limit?
Answer: Look up $10,000 in the CREDIT_LIMIT index to find a collection of row numbers (2, 5, and 7). Use these row numbers to find the corresponding rows in the CUSTOMER table (Brookings Direct, Bargains Galore, and Johnson's Department Store).

Q & A

Question: How would you use the index shown in Figure 7-20 to find every customer represented by sales rep 35?
Answer: Look up 35 in the REP_NUM index to find a collection of row numbers (2, 4, 8, and 9). Use these row numbers to find the corresponding rows in the CUSTOMER table (Brookings Direct, The Everything Shop, Lee's Sport and Appliance, and Deerfield's Four Seasons).

The actual structure of an index is more complicated than what is shown in the figures. Fortunately, you do not have to worry about the details of manipulating and using indexes because the DBMS manages them for you—your only job is to determine the columns on which to build the indexes. Typically, you can create and maintain an index for any column or combination of columns in any table. After creating an index, the DBMS uses it to speed up data retrieval.

As you would expect, the use of any index has advantages and disadvantages. An important advantage was already mentioned: an index makes certain types of retrieval more efficient.

There are two disadvantages when using indexes. First, an index occupies disk space. Using this space for an index, however, is technically unnecessary because any retrieval that you can make using an index also can be made without the index; the index just speeds up the retrieval. The second disadvantage is that the DBMS must update the index whenever corresponding data in the database is updated. Without the index, the DBMS would not need to make these updates. The main question that you must ask when considering whether to create a given index is this: do the benefits derived during retrieval outweigh the additional storage required and the extra processing involved in update operations? In a very large database, you might find that indexes are essential to decrease the time required to retrieve records. In a small database, however, an index might not provide any significant benefits.

You can add and drop indexes as necessary. You can create an index after the database is built; it does not need to be created at the same time as the database. Likewise, when an existing index is no longer necessary, you can drop it.

Creating an Index

Suppose some users at Premiere Products need to display customer records ordered by balance. Other users need to access a customer's name using the customer's number. In addition, some users need to produce a report in which customer records are listed by credit limit in descending order. Within the group of customers having the same credit limit, the customer records must be ordered by name.

Each of the previous requirements is carried out more efficiently when you create the appropriate index. The command used to create an index is **CREATE INDEX**, as illustrated in Example 15.

EXAMPLE 15

Create an index named BALIND on the BALANCE column in the CUSTOMER table. Create an index named REPNAME on the combination of the LAST_NAME and FIRST_NAME columns in the REP table. Create an index named CREDNAME on the combination of the CREDIT_LIMIT and CUSTOMER_NAME columns in the CUSTOMER table, with the credit limits listed in descending order.

The CREATE INDEX command to create the index named BALIND appears in Figure 7-21. The command lists the name of the index and the table name on which the index is to be created. The column on which to create the index—BALANCE—is listed in parentheses.

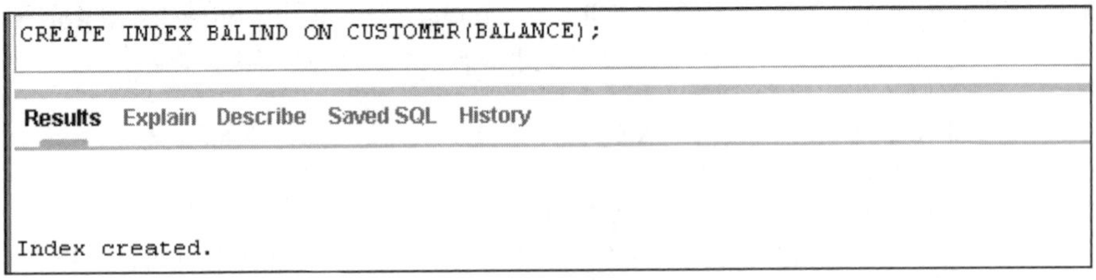

```
CREATE INDEX BALIND ON CUSTOMER(BALANCE);

Results  Explain  Describe  Saved SQL  History

Index created.
```

FIGURE 7-21 Creating the BALIND index on the BALANCE column

The CREATE INDEX command to create the index named REPNAME on the combination of the LAST_NAME and FIRST_NAME columns in the REP table appears in Figure 7-22.

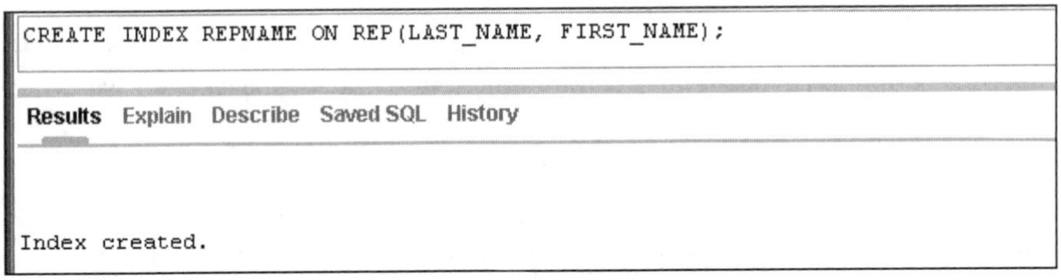

```
CREATE INDEX REPNAME ON REP(LAST_NAME, FIRST_NAME);

Results  Explain  Describe  Saved SQL  History

Index created.
```

FIGURE 7-22 Creating the REPNAME index on the LAST_NAME and FIRST_NAME columns

The CREATE INDEX command to create the index named CREDNAME on the combination of the CREDIT_LIMIT and CUSTOMER_NAME columns in the CUSTOMER table appears in Figure 7-23. When you need to index a column in descending order, the column name is followed by the DESC operator.

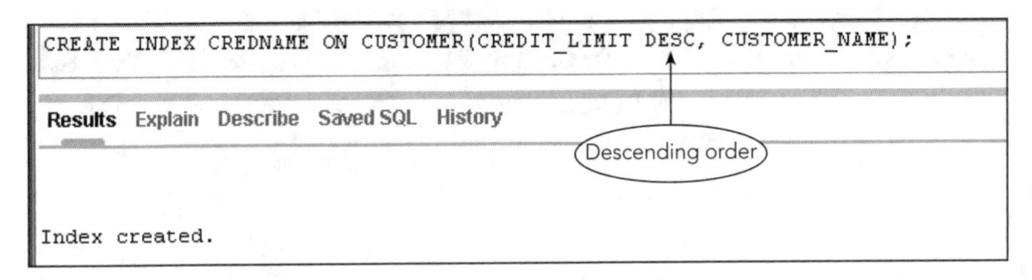

FIGURE 7-23 Creating the CREDNAME index on the CREDIT_LIMIT and CUSTOMER_NAME columns

When customers are listed using the CREDNAME index, the records appear in order by descending credit limit. Within any credit limit, the customers are listed alphabetically by name.

ACCESS USER NOTE

Access supports the creation of indexes in both SQL view and Design view. You can write a CREATE INDEX command in SQL view. You can also open the table containing the column(s) on which you want to create an index in Design view, and click the Indexes button or use the Indexed property to create the index.

Dropping an Index

The command used to drop (delete) an index is **DROP INDEX**, which consists of the words DROP INDEX followed by the name of the index to drop. To delete the CREDNAME index on the CUSTOMER table, for example, the command is:


```
DROP INDEX CREDNAME;
```

The DROP INDEX command permanently deletes the index. CREDNAME was the index the DBMS used when listing customer records in descending order by credit limit order and then by customer name within credit limit. The DBMS still can list customers in this order; however, it cannot do so as efficiently without the index.

SQL SERVER USER NOTE

The SQL Server command to drop an index requires that you qualify the index name. To delete the CREDNAME index on the CUSTOMER table, for example, the command is:

```
DROP INDEX CUSTOMER.CREDNAME;
```

Creating Unique Indexes

When you specify a table's primary key, the DBMS automatically ensures that the values entered in the primary key column(s) are unique. For example, the DBMS rejects an attempt to add a second customer whose number is 148 in the CUSTOMER table because customer 148 already exists. Thus, you do not need to take any special action to make sure that values in the primary key column are unique; the DBMS does it for you.

Occasionally, a nonprimary key column might store unique values. For example, in the REP table, the primary key is REP_NUM. If the REP table also contains a column for Social Security numbers, the values in this column also must be unique because no two people can have the same Social Security number. Because the Social Security number column is not the table's primary key, however, you need to take special action in order for the DBMS to ensure that there are no duplicate values in this column.

To ensure the uniqueness of values in a nonprimary key column, you can create a **unique index** by using the **CREATE UNIQUE INDEX** command. To create a unique index named SSN on the SOC_SEC_NUM column in the REP table, for example, the command is:

```
CREATE UNIQUE INDEX SSN ON REP(SOC_SEC_NUM);
```

This unique index has all the properties of indexes already discussed, along with one additional property: the DBMS rejects any update that causes a duplicate value in the SOC_SEC_NUM column. In this case, the DBMS rejects the addition of a rep whose Social Security number is the same as that of another rep already in the database.

SYSTEM CATALOG

Information about the tables in the database is kept in the **system catalog (catalog)** or the **data dictionary**. This section describes the types of items kept in the catalog and the way in which you can query it to access information about the database structure.

The DBMS automatically maintains the system catalog, which contains several tables. The catalog tables you'll consider in this basic introduction are **SYSTABLES** (information about the tables known to SQL), **SYSCOLUMNS** (information about the columns within these tables), and **SYSVIEWS** (information about the views that have been created). Individual SQL implementations might use different names for these tables. In Oracle, the equivalent tables are named **DBA_TABLES**, **DBA_TAB_COLUMNS**, and **DBA_VIEWS**.

The system catalog is a relational database of its own. Consequently, you can use the same types of queries to retrieve information that you can use to retrieve data in a relational database. You can obtain information about the tables in a relational database, the columns they contain, and the views built on them from the system catalog. The following examples illustrate this process.

NOTE

Most Oracle users need privileges to view system catalog data, so you might not be able to execute these commands. If you are executing the commands shown in the figures, substitute your user name for PRATT to list objects that you own. Your results will differ from those shown in the figures.

ACCESS USER NOTE

In Access, use the Documenter to obtain the information discussed in this section, rather than querying the system catalog.

SQL SERVER USER NOTE

In SQL Server, use stored procedures to obtain the information discussed in this section. To display information about the tables and views in a database, use the sp_tables procedure. For example, the following command displays all the tables and views in the current database:

```
EXEC sp_tables
```

The sp_columns stored procedure displays information about the columns in a particular table. The following command displays the column information for the REP table:

```
EXEC sp_columns REP
```

EXAMPLE 16

List the name of each table for which the owner (creator of the table) is PRATT.

The command to list the table names owned by PRATT is shown in Figure 7-24. The WHERE clause restricts the tables to only those owned by PRATT.

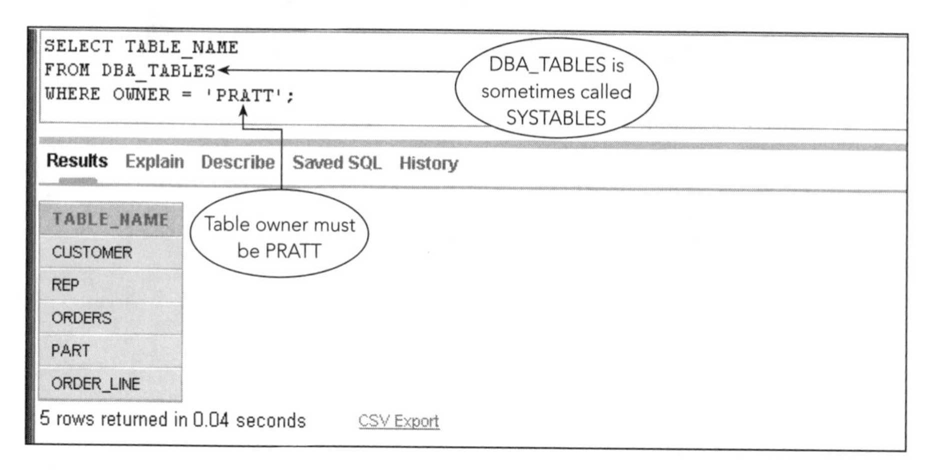

FIGURE 7-24 Tables owned by PRATT

EXAMPLE 17

List the name of each view owned by PRATT.

This command is similar to the command in Example 16. Rather than TABLE_NAME, the column to be selected is named VIEW_NAME. The command appears in Figure 7-25.

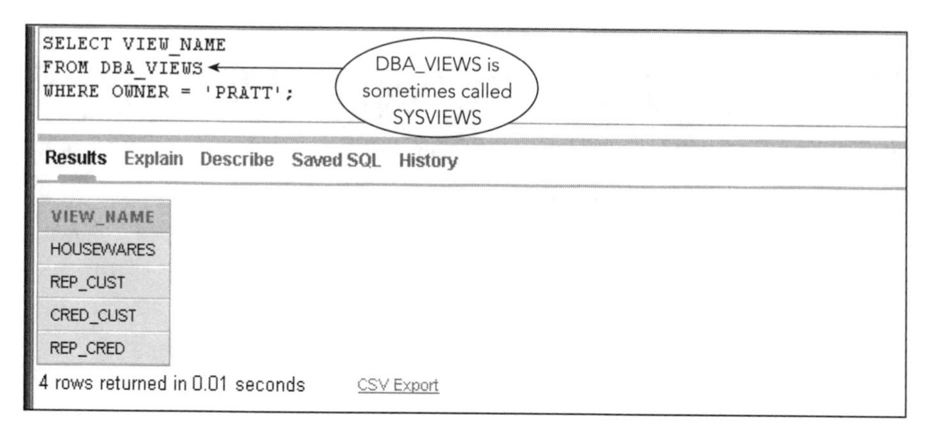

FIGURE 7-25 Views owned by PRATT

EXAMPLE 18

For the CUSTOMER table owned by PRATT, list each column and its data type.

The command for this example appears in Figure 7-26. The columns to select are COLUMN_NAME and DATA_TYPE.

```
SELECT COLUMN_NAME, DATA_TYPE
FROM DBA_TAB_COLUMNS
WHERE OWNER = 'PRATT'
AND TABLE_NAME = 'CUSTOMER';
```

DBA_TAB_COLUMNS is sometimes called SYSCOLUMNS

Results Explain Describe Saved SQL History

COLUMN_NAME	DATA_TYPE
CUSTOMER_NUM	CHAR
CUSTOMER_NAME	CHAR
STREET	CHAR
CITY	CHAR
STATE	CHAR
ZIP	CHAR
BALANCE	NUMBER
CREDIT_LIMIT	NUMBER
REP_NUM	CHAR

9 rows returned in 0.32 seconds CSV Export

FIGURE 7-26 Columns in the CUSTOMER table

EXAMPLE 19

List each table owned by PRATT that contains a column named CUSTOMER_NUM.

As shown in Figure 7-27, the COLUMN_NAME column is used in the WHERE clause to restrict the rows to those in which the column name is CUSTOMER_NUM. (Extra tables generated by Oracle might appear in your list, as shown in Figure 7-27.)

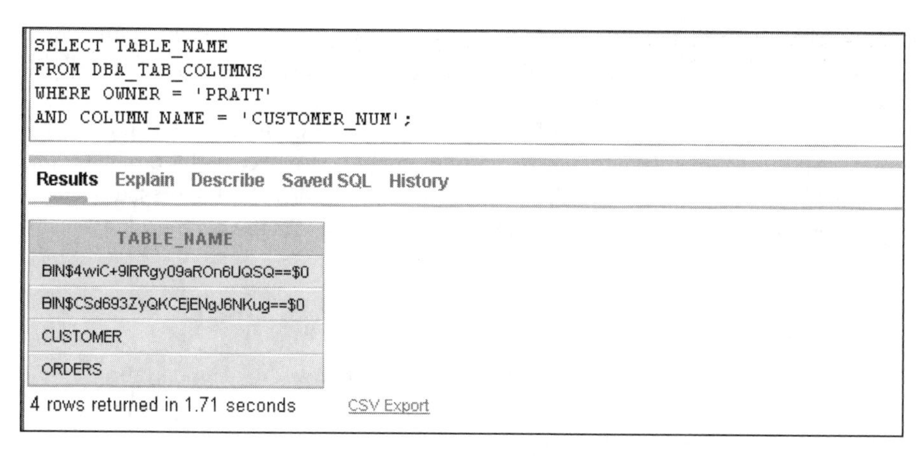
```
SELECT TABLE_NAME
FROM DBA_TAB_COLUMNS
WHERE OWNER = 'PRATT'
AND COLUMN_NAME = 'CUSTOMER_NUM';
```

Results Explain Describe Saved SQL History

TABLE_NAME
BIN$4wiC+9IRRgy09aROn6UQSQ==$0
BIN$CSd693ZyQKCEjENgJ6NKug==$0
CUSTOMER
ORDERS

4 rows returned in 1.71 seconds CSV Export

FIGURE 7-27 Tables owned by PRATT with CUSTOMER_NUM columns

When users create, alter, or drop tables or create or drop indexes, the DBMS updates the system catalog automatically to reflect these changes. Users should not execute SQL queries to update the catalog directly because this might produce inconsistent information. For example, when a user deletes the CUSTOMER_NUM column in the DBA_TAB_COLUMNS table, the DBMS would no longer have any knowledge of this column, which is the CUSTOMER table's primary key, yet all the rows in the CUSTOMER table would still contain a customer number. The DBMS might now treat those customer numbers as names, because as far as the DBMS is concerned, the column named CUSTOMER_NAME is the first column in the CUSTOMER table.

INTEGRITY CONSTRAINTS IN SQL

An **integrity constraint** is a rule for the data in the database. Examples of integrity constraints in the Premiere Products database are as follows:

- A sales rep's number must be unique.
- The sales rep number for a customer must match the number of a sales rep currently in the database. For example, because there is no sales rep number 11, a customer cannot be assigned to sales rep 11.
- Item classes for parts must be AP, HW, or SG because these are the only valid item classes.

If a user enters data in the database that violates any of these integrity constraints, the database develops serious problems. For example, two sales reps with the same number, a customer with a nonexistent sales rep, or a part in a nonexistent item class would compromise the integrity of data in the database. To manage these types of problems, SQL provides **integrity support**, the process of specifying and enforcing integrity constraints for a database. SQL has clauses to support three types of integrity constraints that you can specify within a CREATE TABLE or an ALTER TABLE command. The only difference between these two commands is that an ALTER TABLE command is followed by the word

ADD to indicate that you are adding the constraint to the list of existing constraints. To change an integrity constraint after it has been created, just enter the new constraint, which immediately takes the place of the original.

The types of constraints supported in SQL are primary keys, foreign keys, and legal values. In most cases, you specify a table's primary key when you create the table. To add a primary key after creating a table, you can use the **ADD PRIMARY KEY** clause of the ALTER TABLE command. For example, to indicate that REP_NUM is the primary key for the REP table, the ALTER TABLE command is:


```
ALTER TABLE REP
ADD PRIMARY KEY (REP_NUM);
```

The ADD PRIMARY KEY clause is ADD PRIMARY KEY followed by the column name that makes up the primary key in parentheses. When the primary key contains more than one column, use commas to separate the column names.

ACCESS USER NOTE

To specify a table's primary key in Access, open the table in Design view, select the column(s) that make up the primary key, and then click the Primary Key button on the Ribbon or the toolbar.

A **foreign key** is a column in one table whose values match the primary key in another table. (One example is the CUSTOMER_NUM column in the ORDERS table. Values in this column are required to match those of the primary key in the CUSTOMER table.)

EXAMPLE 20

Specify the CUSTOMER_NUM column in the ORDERS table as a foreign key that must match the CUSTOMER table.

When a table contains a foreign key, you identify it using the **ADD FOREIGN KEY** clause of the ALTER TABLE command. In this clause, you specify the column that is a foreign key *and* the table it matches. The general form for assigning a foreign key is ADD FOREIGN KEY, the column name(s) of the foreign key, the REFERENCES clause, and then the table name that the foreign key must match, as shown in Figure 7-28.

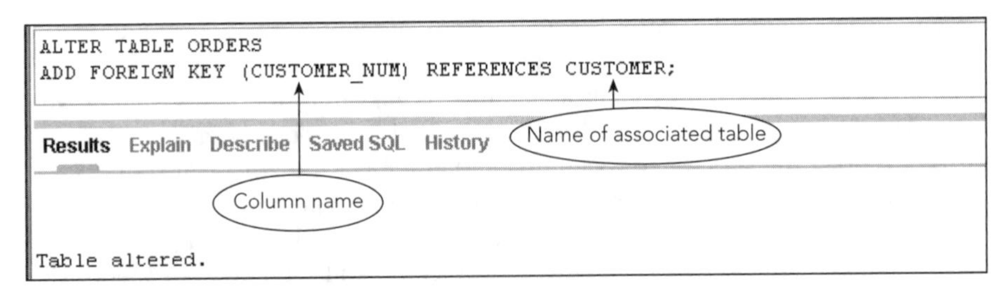

FIGURE 7-28 Adding a foreign key to an existing table

ACCESS USER NOTE

To specify a foreign key in Access, open the Relationships window, relate the corresponding tables on the matching column, and then select the option to enforce referential integrity.

After creating a foreign key, the DBMS rejects any update that violates the foreign key constraint. For example, the DBMS rejects the INSERT command shown in Figure 7-29 because it attempts to add an order for which the customer number (850) does not match any customer in the CUSTOMER table.

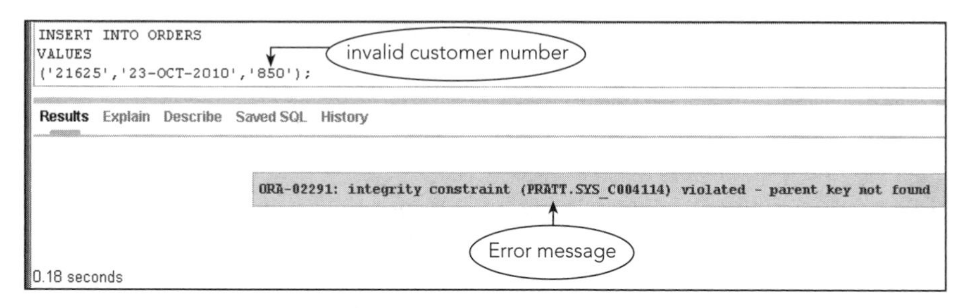

FIGURE 7-29 Violating a foreign key constraint when adding a row

The DBMS also rejects the DELETE command in Figure 7-30 because it attempts to delete customer number 148; rows in the ORDERS table for which the customer number is 148 would no longer match any row in the CUSTOMER table.

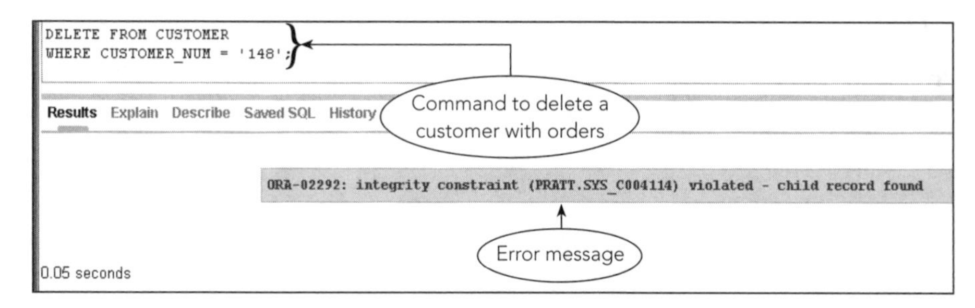

FIGURE 7-30 Violating a foreign key constraint when deleting a row

Note that the error messages shown in Figures 7-29 and 7-30 include the words "parent" and "child." When you specify a foreign key, the table containing the foreign key is the **child**, and the table referenced by the foreign key is the **parent**. For example, the CUSTOMER_NUM column in the ORDERS table is a foreign key that references the CUSTOMER table. For this foreign key, the CUSTOMER table is the parent, and the ORDERS table is the child. The error message shown in Figure 7-29 indicates that there is no parent for the order (there is no customer number 850). The error message shown in Figure 7-30 indicates that there are child records (rows) for customer 148 (customer 148 has orders). The DBMS rejects both updates because they violate referential integrity.

EXAMPLE 21

Specify the valid item classes for the PART table as AP, HW, and SG.

You use the **CHECK** clause of the ALTER TABLE command to ensure that only legal values satisfying a particular condition are allowed in a given column. The general form of the CHECK clause is the word CHECK followed by a condition. If a user enters data that violates the condition, the DBMS rejects the update automatically. For example, to ensure that the only legal values for item class are AP, HW, or SG, use one of the following versions of the CHECK clause:


```
CHECK (CLASS IN ('AP', 'HW', 'SG') )
```

or

```
CHECK (CLASS = 'AP' OR CLASS = 'HW' OR CLASS = 'SG')
```

The ALTER TABLE command shown in Figure 7-31 uses the first version of the CHECK clause.

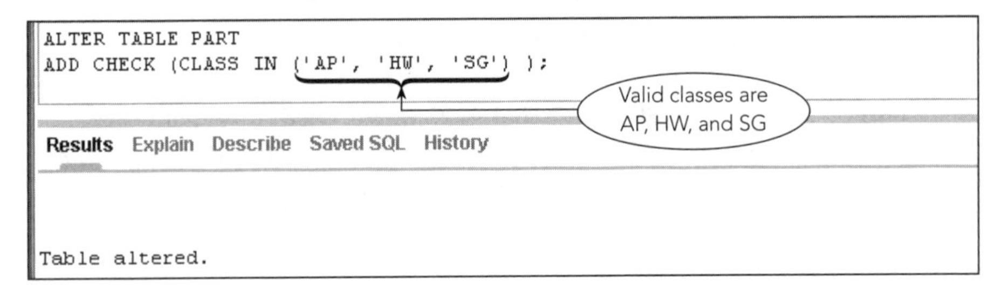

FIGURE 7-31 Adding an integrity constraint to an existing table

Now the DBMS will reject the update shown in Figure 7-32 because the command attempts to change the item class to XX, which is an illegal value.

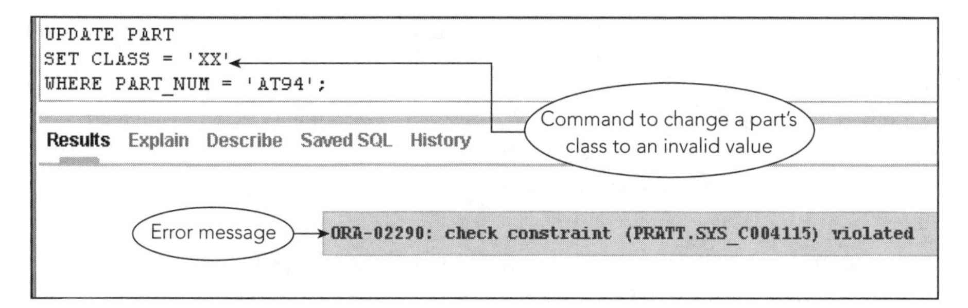

FIGURE 7-32 Update that violates an integrity constraint

ACCESS USER NOTE

Access does not support the CHECK clause. To specify a validation rule in Access, open the table in Design view, and then enter an expression in the column's Validation Rule property to limit the values that users can enter into the column.

Chapter Summary

- A view contains data that is derived from existing base tables when users attempt to access the view.
- To create a view, use the CREATE VIEW command, which includes a defining query that describes the portion of the database included in the view. When a user retrieves data from the view, the DBMS merges the query entered by the user with the defining query and produces the query that the DBMS actually executes.
- Views provide data independence, allow database access control, and simplify the database structure for users.
- You cannot update views that involve statistics and views with joins of nonprimary key columns. Updates for these types of views must be made in the base table.
- Use the DROP VIEW command to delete a view.
- Use the GRANT command to give users access privileges to data in the database.
- Use the REVOKE command to terminate previously granted privileges.
- You can create and use an index to make data retrieval more efficient. Use the CREATE INDEX command to create an index. Use the CREATE UNIQUE INDEX command to enforce a rule so only unique values are allowed in a nonprimary key column.
- Use the DROP INDEX command to delete an index.
- The DBMS, not the user, chooses which index to use to accomplish a given task.
- The DBMS maintains information about the tables, columns, indexes, and other system elements in the system catalog (catalog) or data dictionary. Information about tables is kept in the SYSTABLES table, information about columns is kept in the SYSCOLUMNS table, and information about views is kept in the SYSVIEWS table. In Oracle, these same tables are named DBA_TABLES, DBA_TAB_COLUMNS, and DBA_VIEWS.
- Use the SELECT command to obtain information from the system catalog. The DBMS updates the system catalog automatically whenever changes are made to the database. In Access, use the Documenter to obtain information about the database objects. SQL Server uses stored procedures to obtain information from the system catalog.
- Integrity constraints are rules that the data in the database must follow to ensure that only legal values are accepted in specified columns and that primary and foreign key values match between tables. To specify a general integrity constraint, use the CHECK clause. You usually specify primary key constraints when you create a table, but you can specify them later using the ADD PRIMARY KEY clause. To specify a foreign key, use the ADD FOREIGN KEY clause.

Key Terms

ADD FOREIGN KEY
ADD PRIMARY KEY
base table
catalog
CHECK
child
CREATE INDEX
CREATE UNIQUE INDEX
CREATE VIEW
data dictionary

database administration
database administrator
DBA_TAB_COLUMNS
DBA_TABLES
DBA_VIEWS
defining query
DROP INDEX
DROP VIEW
foreign key
GRANT
index
integrity constraint
integrity support
parent
REFERENCES
REVOKE
row-and-column subset view
security
SYSCOLUMNS
SYSTABLES
system catalog
SYSVIEWS
unique index
view
WITH GRANT OPTION

Review Questions

1. What is a view?
2. Which command creates a view?
3. What is a defining query?
4. What happens when a user retrieves data from a view?
5. What are three advantages of using views?
6. Which types of views cannot be updated?
7. Which command deletes a view?
8. Which command gives users access privileges to various portions of the database?
9. Which command terminates previously granted privileges?
10. What is the purpose of an index?
11. How do you create an index? How do you create a unique index? What is the difference between an index and a unique index?
12. Which command deletes an index?
13. Does the DBMS or the user make the choice of which index to use to accomplish a given task?
14. Describe the information the DBMS maintains in the system catalog. What are the generic names for three tables in the catalog and their corresponding names in Oracle?
15. Use your favorite Web browser and search engine to find information about a data dictionary. Write a one-page paper that describes other types of information that can be stored in a data dictionary. Be sure to cite the URLs that you use.
16. Which command do you use to obtain information from the system catalog in Oracle?
17. How is the system catalog updated?
18. What are integrity constraints?
19. How do you specify a general integrity constraint?

20. When would you usually specify primary key constraints? Can you specify them after creating a table? How?
21. How do you specify a foreign key in Oracle?
22. Use your favorite Web browser and search engine to find information about referential integrity. Write two or three paragraphs that describe what referential integrity is and include an example of how referential integrity is used in the Premiere Products database. Be sure to cite the URLs that you use.

Exercises

Premiere Products

Use SQL to make the following changes to the Premiere Products database (see Figure 1-2 in Chapter 1). After each change, execute an appropriate query to show that the change was made correctly. If directed to do so by your instructor, use the information provided with the Chapter 3 Exercises to print your output. For any exercises that use commands not supported by your version of SQL, write the command to accomplish the task.

1. Create a view named MAJOR_CUSTOMER. It consists of the customer number, name, balance, credit limit, and rep number for every customer whose credit limit is $10,000 or less.
   a. Write and execute the CREATE VIEW command to create the MAJOR_CUSTOMER view.
   b. Write and execute the command to retrieve the customer number and name of each customer in the MAJOR_CUSTOMER view with a balance that exceeds the credit limit.
   c. Write and execute the query that the DBMS actually executes.
   d. Does updating the database through this view create any problems? If so, what are they? If not, why not?
2. Create a view named PART_ORDER. It consists of the part number, description, price, order number, order date, number ordered, and quoted price for all order lines currently on file.
   a. Write and execute the CREATE VIEW command to create the PART_ORDER view.
   b. Write and execute the command to retrieve the part number, description, order number, and quoted price for all orders in the PART_ORDER view for parts with quoted prices that exceed $100.
   c. Write and execute the query that the DBMS actually executes.
   d. Does updating the database through this view create any problems? If so, what are they? If not, why not?
3. Create a view named ORDER_TOTAL. It consists of the order number and order total for each order currently on file. (The order total is the sum of the number of units ordered multiplied by the quoted price on each order line for each order.) Sort the rows by order number. Use TOTAL_AMOUNT as the name for the order total.
   a. Write and execute the CREATE VIEW command to create the ORDER_TOTAL view.
   b. Write and execute the command to retrieve the order number and order total for only those orders totaling more than $1,000.

c. Write and execute the query that the DBMS actually executes.

d. Does updating the database through this view create any problems? If so, what are they? If not, why not?

4. Write, but do not execute, the commands to grant the following privileges:
   a. User Ashton must be able to retrieve data from the PART table.
   b. Users Kelly and Morgan must be able to add new orders and order lines.
   c. User James must be able to change the price for all parts.
   d. User Danielson must be able to delete customers.
   e. All users must be able to retrieve each customer's number, name, street, city, state, and zip code.
   f. User Perez must be able to create an index on the ORDERS table.
   g. User Washington must be able to change the structure of the PART table.
   h. User Grinstead must have all privileges on the ORDERS table.
5. Write, but do not execute, the command to revoke all privileges from user Ashton.
6. Perform the following tasks:
   a. Create an index named PART_INDEX1 on the PART_NUM column in the ORDER_LINE table.
   b. Create an index named PART_INDEX2 on the CLASS column in the PART table.
   c. Create an index named PART_INDEX3 on the CLASS and WAREHOUSE columns in the PART table.
   d. Create an index named PART_INDEX4 on the CLASS and WAREHOUSE columns in the PART table. List item classes in descending order.
7. Delete the index named PART_INDEX3.
8. Write the commands to obtain the following information from the system catalog. Do not execute these commands unless your instructor asks you to do so.
   a. List every table that contains a column named CUSTOMER_NUM.
   b. List every column in the PART table and its associated data type.
9. Add the ORDER_NUM column as a foreign key in the ORDER_LINE table.
10. Ensure that the only values entered into the CREDIT_LIMIT column are 5000, 7500, 10000, and 15000.

Henry Books

Use SQL to make the following changes to the Henry Books database (Figures 1-4 through 1-7 in Chapter 1). After each change, execute an appropriate query to show that the change was made correctly. If directed to do so by your instructor, use the information provided with the Chapter 3 Exercises to print your output. For any exercises that use commands not supported by your version of SQL, write the command to accomplish the task.

1. Create a view named PLUME. It consists of the book code, title, type, and price for every book published by the publisher whose code is PL.
   a. Write and execute the CREATE VIEW command to create the PLUME view.
   b. Write and execute the command to retrieve the book code, title, and price for every book with a price of less than $13.
   c. Write and execute the query that the DBMS actually executes.
   d. Does updating the database through this view create any problems? If so, what are they? If not, why not?

2. Create a view named NONPAPERBACK. It consists of the book code, title, publisher name, and price for every book that is not available in paperback.
   a. Write and execute the CREATE VIEW command to create the NONPAPERBACK view.
   b. Write and execute the command to retrieve the book title, publisher name, and price for every book in the NONPAPERBACK view with a price of less than $20.
   c. Write and execute the query that the DBMS actually executes.
   d. Does updating the database through this view create any problems? If so, what are they? If not, why not?
3. Create a view named BOOK_INVENTORY. It consists of the branch number and the total number of books on hand for each branch. Use UNITS as the name for the count of books on hand. Group and order the rows by branch number.
   a. Write and execute the CREATE VIEW command to create the BOOK_INVENTORY view.
   b. Write and execute the command to retrieve the branch number and units for each branch having more than 25 books on hand.
   c. Write and execute the query that the DBMS actually executes.
   d. Does updating the database through this view create any problems? If so, what are they? If not, why not?
4. Write, but do not execute, the commands to grant the following privileges:
   a. User Rodriquez must be able to retrieve data from the BOOK table.
   b. Users Gomez and Liston must be able to add new books and publishers to the database.
   c. Users Andrews and Zimmer must be able to change the price of any book.
   d. All users must be able to retrieve the book title, book code, and book price for every book.
   e. User Golden must be able to add and delete publishers.
   f. User Andrews must be able to create an index for the BOOK table.
   g. Users Andrews and Golden must be able to change the structure of the AUTHOR table.
   h. User Golden must have all privileges on the BRANCH, BOOK, and INVENTORY tables.
5. Write, but do not execute, the command to revoke all privileges from user Andrews.

6. Create the following indexes:
   a. Create an index named BOOK_INDEX1 on the TITLE column in the BOOK table.
   b. Create an index named BOOK_INDEX2 on the TYPE column in the BOOK table.
   c. Create an index named BOOK_INDEX3 on the CITY and PUBLISHER_NAME columns in the PUBLISHER table.
7. Delete the index named BOOK_INDEX3.
8. Write the commands to obtain the following information from the system catalog. Do not execute these commands unless your instructor asks you to do so.
   a. List every column in the PUBLISHER table and its associated data type.
   b. List every table that contains a column named PUBLISHER_CODE.
   c. List the table name, column name, and data type for the columns named BOOK_CODE, TITLE, and PRICE. Order the results by table name within column name. (That is, column name is the major sort key and table name is the minor sort key.)
9. Add the PUBLISHER_CODE column as a foreign key in the BOOK table.
10. Ensure that the PAPERBACK column in the BOOK table can accept only values of Y or N.

Alexamara Marina Group

Use SQL to make the following changes to the Alexamara Marina Group database (Figures 1-8 through 1-12 in Chapter 1). After each change, execute an appropriate query to show that the change was made correctly. If directed to do so by your instructor, use the information provided with the Chapter 3 Exercises to print your output. For any exercises that use commands not supported by your version of SQL, write the command to accomplish the task.

1. Create a view named LARGE_SLIP. It consists of the marina number, slip number, rental fee, boat name, and owner number for every slip whose length is 40 feet.
   a. Write and execute the CREATE VIEW command to create the LARGE_SLIP view.
   b. Write and execute the command to retrieve the marina number, slip number, rental fee, and boat name for every slip with a rental fee of $3,800 or more.
   c. Write and execute the query that the DBMS actually executes.
   d. Does updating the database through this view create any problems? If so, what are they? If not, why not?
2. Create a view named RAY_4025. It consists of the marina number, slip number, length, rental fee, boat name, and owner's last name for every slip in which the boat type is Ray 4025.
   a. Write and execute the CREATE VIEW command to create the RAY_4025 view.
   b. Write and execute the command to retrieve the marina number, slip number, rental fee, boat name, and owner's last name for every slip in the RAY_4025 view with a rental fee of less than $4,000.

c. Write and execute the query that the DBMS actually executes.

d. Does updating the database through this view create any problems? If so, what are they? If not, why not?

3. Create a view named SLIP_FEES. It consists of two columns: the first is the slip length, and the second is the average fee for all slips in the MARINA_SLIP table that have that length. Use AVERAGE_FEE as the name for the average fee. Group and order the rows by slip length.

   a. Write and execute the CREATE VIEW command to create the SLIP_FEES view.

   b. Write and execute the command to retrieve the slip length and average fee for each length for which the average fee is less than $3,500.

   c. Write and execute the query that the DBMS actually executes.

   d. Does updating the database through this view create any problems? If so, what are they? If not, why not?

4. Write, but do not execute, the commands to grant the following privileges:

   a. User Oliver must be able to retrieve data from the MARINA_SLIP table.

   b. Users Crandall and Perez must be able to add new owners and slips to the database.

   c. Users Johnson and Klein must be able to change the rental fee of any slip.

   d. All users must be able to retrieve the length, boat name, and owner number for every slip.

   e. User Klein must be able to add and delete service categories.

   f. User Adams must be able to create an index on the SERVICE_REQUEST table.

   g. Users Adams and Klein must be able to change the structure of the MARINA_SLIP table.

   h. User Klein must have all privileges on the MARINA, OWNER, and MARINA_SLIP tables.

5. Write, but do not execute, the command to revoke all privileges from user Adams.

6. Create the following indexes:

   a. Create an index named BOAT_INDEX1 on the OWNER_NUM column in the MARINA_SLIP table.

   b. Create an index named BOAT_INDEX2 on the BOAT_NAME column in the MARINA_SLIP table.

   c. Create an index named BOAT_INDEX3 on the LENGTH and BOAT_NAME columns in the MARINA_SLIP table. List the lengths in descending order.

7. Delete the index named BOAT_INDEX3; it is no longer necessary.

8. Write the commands to obtain the following information from the system catalog. Do not execute these commands unless your instructor specifically asks you to do so.

   a. List every column in the MARINA_SLIP table and its associated data type.

   b. List every table and view that contains a column named MARINA_NUM.

9. Add the OWNER_NUM column as a foreign key in the MARINA_SLIP table.

10. Ensure that the LENGTH column in the MARINA_SLIP table can accept only values of 25, 30, or 40.

CHAPTER 8

SQL FUNCTIONS AND PROCEDURES

LEARNING OBJECTIVES

Objectives

- Understand how to use functions in queries
- Use the UPPER and LOWER functions with character data
- Use the ROUND and FLOOR functions with numeric data
- Add a specific number of months or days to a date
- Calculate the number of days between two dates
- Use concatenation in a query
- Embed SQL commands in PL/SQL and T-SQL procedures
- Retrieve single rows using embedded SQL
- Update a table using embedded INSERT, UPDATE, and DELETE commands
- Use cursors to retrieve multiple rows in embedded SQL
- Manage errors in procedures containing embedded SQL commands
- Use SQL in a language that does not support embedded SQL commands
- Use triggers

INTRODUCTION

You already have used functions that apply to groups (such as SUM and AVG). In this chapter, you will learn to use functions that apply to values in individual rows. Specifically, you will see how to use functions with characters or text, numbers, and dates. You will learn how to concatenate values in a query. You will embed SQL commands in PL/SQL and T-SQL procedures to retrieve rows and update data. You will examine the different ways to manage errors in procedures. Finally, you will learn how to create and use cursors and triggers.

USING SQL IN A PROGRAMMING ENVIRONMENT

SQL is a powerful **nonprocedural language** in which you communicate tasks to the computer using simple commands. As in other nonprocedural languages, you can accomplish many tasks using a single command. Although SQL and other nonprocedural languages are well-equipped to store and query data, sometimes you might need to complete tasks that are beyond the capabilities of SQL. In such cases, you need to use a procedural language.

A **procedural language** is one in which you must give the computer the step-by-step process for accomplishing a task. **PL/SQL**, which was developed by Oracle as an extension of SQL, is an example of a procedural language. This chapter uses PL/SQL to illustrate how to use SQL in a programming environment by **embedding** SQL commands in another language. The examples in this chapter illustrate how to use embedded SQL commands to retrieve a single row, insert new rows, update and delete existing rows, and retrieve multiple rows. In the process, you will create stored procedures that are saved and are available for use at any time.

T-SQL, which stands for **Transact-SQL**, is another extension of SQL. T-SQL is the procedural language that SQL Server uses. You can perform tasks, such as retrieving a single row, inserting new rows, and retrieving multiple rows, using T-SQL in SQL Server. Although the language syntax is slightly different in T-SQL when compared to PL/SQL, the functionality and the results are the same.

You cannot embed SQL commands in Access programs the way you can in PL/SQL and T-SQL. There are ways to use the commands, however, as you'll learn later in this chapter.

NOTE

This chapter assumes that you have some programming background and does not cover programming basics. To understand the first part of this chapter, you should be familiar with variables, declaring variables, and creating procedural code, including IF statements and loops. To understand the Access section at the end of the chapter, you should be familiar with Function and Sub procedures, and the process for sequentially accessing all records in a recordset, such as using a loop to process all the records in a table.

ACCESS USER NOTE

If you are using Access, you will not be able to complete the material in this chapter that deals with PL/SQL and T-SQL procedures. Be sure to read this information so you will understand these important concepts. You will, however, be able to complete the steps in the "Using SQL in Microsoft Access" section.

USING FUNCTIONS

You already have used aggregate functions to perform calculations based on groups of records. For example, SUM(BALANCE) calculates the sum of the balances on all records that satisfy the condition in the WHERE clause. When you use a GROUP BY clause, the DBMS will calculate the sum for each record in a group.

SQL also includes functions that affect single records. Some functions affect character data and others let you manipulate numeric data. The supported SQL functions vary between SQL implementations. This section will illustrate some common functions. For additional information about the functions your SQL implementation supports, consult the program's documentation.

Character Functions

SQL includes several functions that affect character data. Example 1 illustrates the use of the UPPER function.

EXAMPLE 1

List the rep number and last name for each sales rep. Display the last name in uppercase letters.

The **UPPER** function displays a value in uppercase letters; for example, the function UPPER(LAST_NAME) displays the last name Kaiser as KAISER. (Note that the UPPER function simply displays the last name in uppercase letters; it does not change the last name stored in the table to uppercase letters.) The item in parentheses (LAST_NAME) is called the **argument** for the function. The value produced by the function is the result of displaying all lowercase letters in the value stored in the LAST_NAME column in uppercase letters. The query and its results are shown in Figure 8-1.

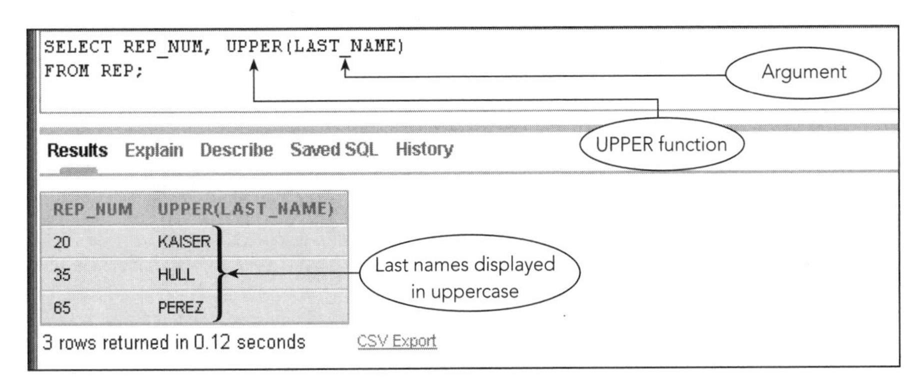

FIGURE 8-1 Using the UPPER function to display character data in uppercase letters

You can use functions in WHERE clauses as well. For example, the condition UPPER(LAST_NAME) = 'KAISER' would be true for names like Kaiser, KAISER, and KaIsER, because the result of applying the UPPER function to any of these values would result in the value KAISER.

To display a value in lowercase letters, you can use the **LOWER** function. SQL Server supports both the UPPER and LOWER function.

ACCESS USER NOTE

In Access, the UCASE() function displays a value in uppercase letters and the LCASE() function displays a value in lowercase letters. For example, if the value stored in the LAST_NAME column is Kaiser, UCASE(LAST_NAME) would result in the value KAISER and LCASE(LAST_NAME) would result in the value kaiser.

Number Functions

SQL also includes functions that affect numeric data. The **ROUND** function, which rounds values to a specified number of decimal places, is illustrated in Example 2.

EXAMPLE 2

List the part number and price for all parts. Round the price to the nearest whole dollar amount.

A function can have more than one argument. The ROUND function, which rounds a numeric value to a desired number of decimal places, has two arguments. The first argument is the value to be rounded; the second argument indicates the number of decimal places to which to round the result. For example, ROUND(PRICE,0) will round the values in the PRICE column to zero decimal places (a whole number). If a price is 24.95, the result will be 25. If the price is 24.25, on the other hand, the result will be 24. Figure 8-2 shows the query and results to round values in the PRICE column to zero decimal places. The computed column ROUND(PRICE,0) is named ROUNDED_PRICE.

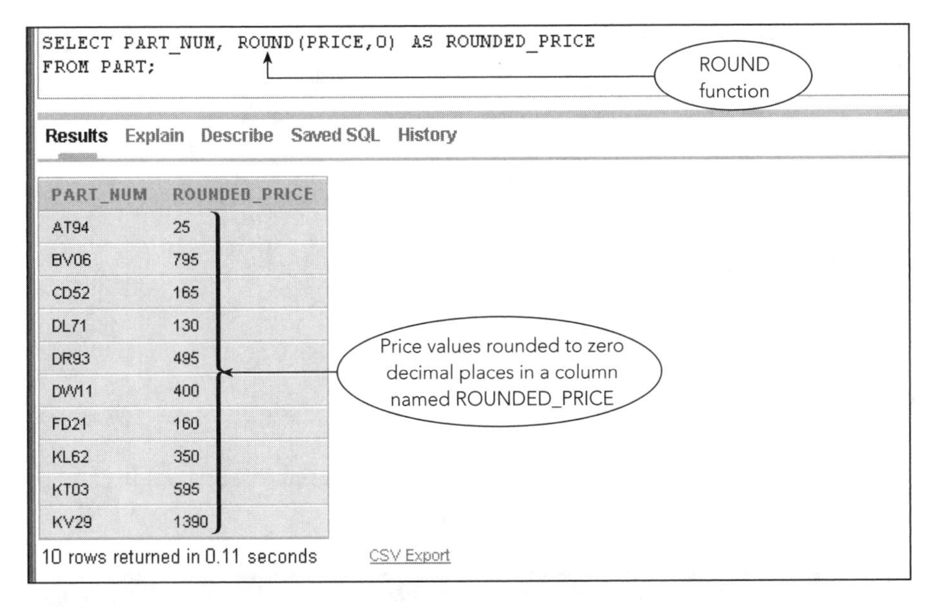

FIGURE 8-2 Using the ROUND function to round numeric values

Rather than rounding (using the ROUND function), you might need to truncate (remove) everything to the right of the decimal point. To do so, use the **FLOOR** function, which has only one argument. If a price is 24.95, for example, ROUND(PRICE,0) would result in 25, whereas FLOOR(PRICE) would result in 24. SQL Server supports both the ROUND and the FLOOR functions. Microsoft Access supports only the ROUND function.

Working with Dates

SQL uses functions and calculations for manipulating dates. To add a specific number of months to a date, you can use the **ADD_MONTHS** function as illustrated in Example 3.

EXAMPLE 3

For each order, list the order number and the date that is two months after the order date. Name this date FUTURE_DATE.

The ADD_MONTHS function has two arguments. The first argument is the date to which you want to add a specific number of months, and the second argument is the number of months. To add two months to the order date, for example, the expression is ADD_MONTHS(ORDER_DATE,2) as illustrated in Figure 8-3.

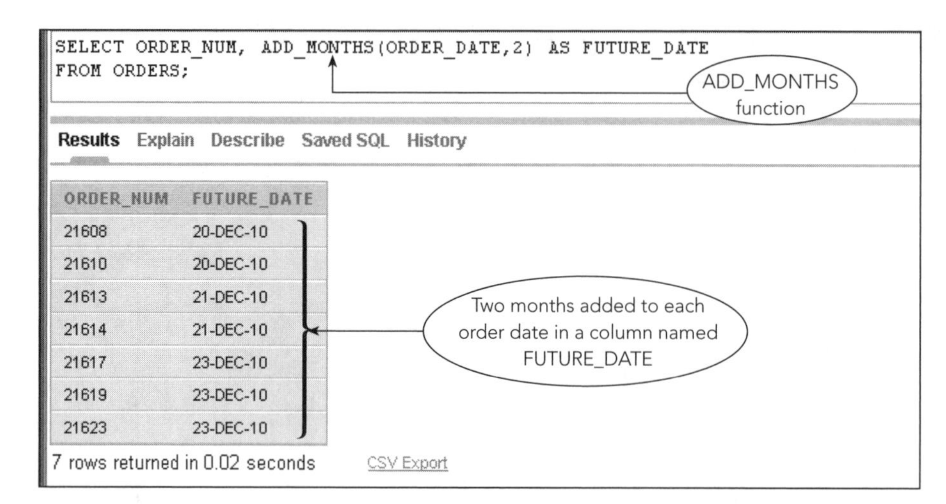

FIGURE 8-3 Using the ADD_MONTHS function to add months to a date

ACCESS USER NOTE

To add a number of months to a date in Access, use the DATEADD() function, which has three arguments. The first argument includes the interval of time to be added; the letter "m" indicates that months will be added. The second argument includes the number of intervals to be added. The third argument includes the date to be manipulated. For example, to add two months to the dates stored in the ORDER_DATE column, the appropriate function would be DATEADD("m", 2, ORDER_DATE).

SQL SERVER USER NOTE

To add a number of months to a date in SQL Server, use the DATEADD() function, which has three arguments. The first argument includes the interval of time to be added; the letter "m" indicates that months will be added. The second argument includes the number of intervals to be added. The third argument includes the date to be manipulated. For example, to add two months to the dates stored in the ORDER_DATE column, the appropriate function would be DATEADD("m", 2, ORDER_DATE).

EXAMPLE 4

For each order, list the order number and the date that is seven days after the order date. Name this date FUTURE_DATE.

To add a specific number of days to a date, you do not need a function. You can add the number of days to the order date as illustrated in Figure 8-4. (You can also subtract dates in the same way.) This method works in Oracle, Access, and SQL Server.

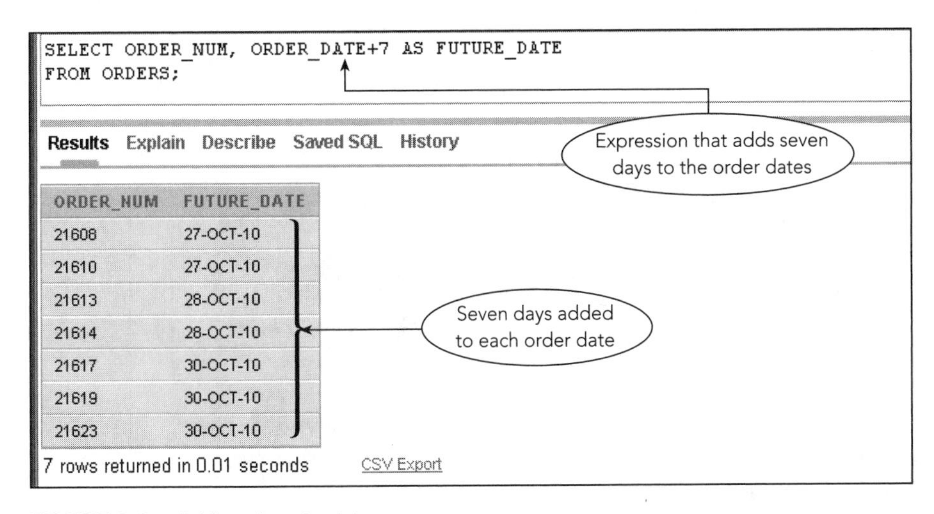

FIGURE 8-4 Adding days to dates

EXAMPLE 5

For each order, list the order number, today's date, the order date, and the number of days between the order date and today's date. Name today's date TODAYS_DATE and name the number of days between the order date and today's date DAYS_PAST.

You can use the **SYSDATE** function to obtain today's date, as shown in Figure 8-5. The command in the figure uses SYSDATE to display today's date and also uses SYSDATE in a computation to determine the number of days between the order date and today's date. The values for DAYS_PAST include decimal places. You could remove these decimal places by using the ROUND or FLOOR functions, if desired.

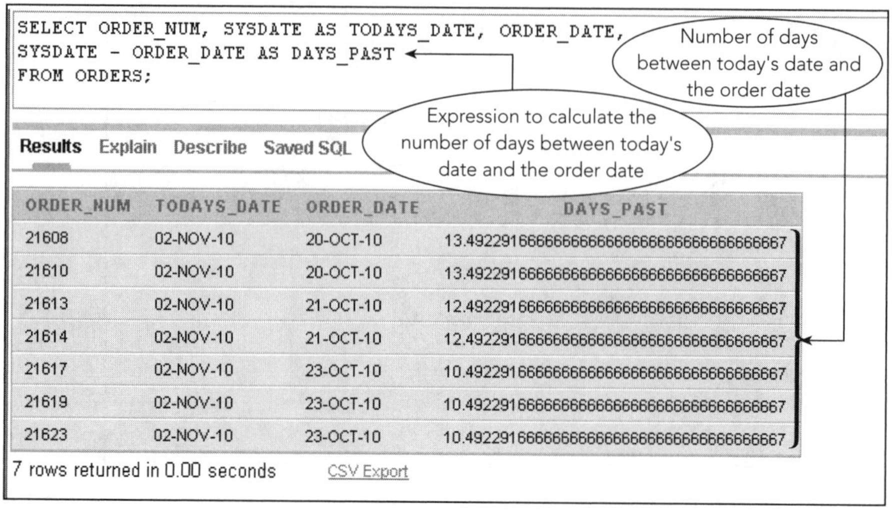

SELECT ORDER_NUM, SYSDATE AS TODAYS_DATE, ORDER_DATE,
SYSDATE - ORDER_DATE AS DAYS_PAST
FROM ORDERS;
Number of days between today's date and the order date
Expression to calculate the number of days between today's date and the order date
Results Explain Describe Saved SQL
ORDER_NUM TODAYS_DATE ORDER_DATE DAYS_PAST
21608 02-NOV-10 20-OCT-10 13.49229166666666666666666666666666667
21610 02-NOV-10 20-OCT-10 13.49229166666666666666666666666666667
21613 02-NOV-10 21-OCT-10 12.49229166666666666666666666666666667
21614 02-NOV-10 21-OCT-10 12.49229166666666666666666666666666667
21617 02-NOV-10 23-OCT-10 10.49229166666666666666666666666666667
21619 02-NOV-10 23-OCT-10 10.49229166666666666666666666666666667
21623 02-NOV-10 23-OCT-10 10.49229166666666666666666666666666667
7 rows returned in 0.00 seconds
CSV Export

FIGURE 8-5 Calculating the number of days between two dates

ACCESS USER NOTE

In Access, use the DATE() function to obtain today's date, rather than SYSDATE. The DATE() function has no arguments, so you would write DATE() in place of SYSDATE.

SQL SERVER USER NOTE

In SQL Server, use the GETDATE() function to obtain today's date, rather than SYSDATE. The GETDATE() function has no arguments, so you would write GETDATE() in place of SYSDATE.

CONCATENATING COLUMNS

Sometimes you need to **concatenate**, or combine, two or more character columns into a single expression when displaying them in a query; the process is called **concatenation**. To concatenate columns, you type two vertical lines (||) between the column names, as illustrated in Example 6.

EXAMPLE 6

List the number and name of each sales rep. Concatenate the FIRST_NAME and LAST_NAME columns into a single value, with a space separating the first and last names.

To concatenate the FIRST_NAME and LAST_NAME columns, the expression is FIRST_NAME||LAST_NAME. When the first name doesn't include sufficient characters to fill the width of the column (as determined by the number of characters specified in the CREATE TABLE command), SQL inserts extra spaces. For example, when the

FIRST_NAME column is 12 characters wide, the first name is Mary, and the last name is Johnson, the expression FIRST_NAME||LAST_NAME appears as Mary, followed by eight spaces, and then Johnson. To remove the extra spaces following the first name value, you use the **RTRIM** (right trim) function. When you apply this function to the value in a column, SQL displays the original value and removes any spaces inserted at the end of the value. Figure 8-6 shows the query and output with the extra spaces removed. For sales rep 20, for example, this command trims the first name to "Valerie," concatenates it with a single space, and then concatenates the last name "Kaiser."

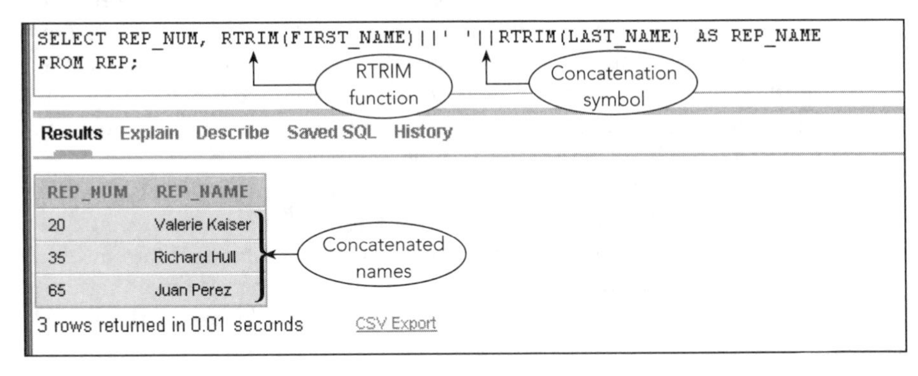

FIGURE 8-6 Concatenating two columns and using the RTRIM function

Q & A

Question: Why is it necessary to insert a single space character in single quotation marks in the query?
Answer: Without the space character, there would be no space between the first and last names. The name of sales rep 20, for example, would be displayed as "ValerieKaiser."

ACCESS USER NOTE

In Access, use the & symbol to concatenate columns. It is not necessary to trim the columns because Access will trim them automatically. The corresponding query in Access is:

```
SELECT REP_NUM, FIRST_NAME&' '&LAST_NAME
FROM REP;
```

SQL SERVER USER NOTE

In SQL Server, use the + symbol to concatenate columns. The corresponding query in SQL Server is:

```
SELECT REP_NUM, RTRIM(FIRST_NAME)+' '+RTRIM(LAST_NAME)
FROM REP;
```

STORED PROCEDURES

In a **client/server system**, the database is stored on a computer called the **server** and users access the database through clients. A **client** is a computer that is connected to a network and has access through the server to the database. Every time a user executes a query, the DBMS must determine the best way to process the query and provide the results. For example, the DBMS must determine which indexes are available and whether it can use those indexes to make the processing of the query more efficient.

When you anticipate running a particular query often, you can improve overall performance by saving the query in a file called a **stored procedure**. The stored procedure is placed on the server. The DBMS compiles the stored procedure (translating it into machine code) and creates an execution plan, which is the most efficient way of obtaining the results. From that point on, users execute the compiled, optimized code in the stored procedure.

Another reason for saving a query as a stored procedure, even when you are not working in a client/server system, is convenience. Rather than retyping the entire query each time you need it, you can use the stored procedure. For example, suppose you frequently execute a query that selects a sales rep with a given number and then displays the concatenation of the first name and last name of the sales rep. Instead of typing the query each time you want to display a sales rep's name, you can store the query in a stored procedure. You would then only need to run the stored procedure when you want to display a sales rep's name.

ACCESS USER NOTE

Although Access does not support stored procedures, you can achieve some of the same convenience by creating a parameter query that prompts the user for the arguments you would otherwise use in a stored procedure.

In Oracle, you create stored procedures using a language called PL/SQL. You create and save the procedures as script files.

Retrieving a Single Row and Column

Example 7 illustrates using a stored procedure to retrieve a single row and column from a table.

EXAMPLE 7

Write a PL/SQL procedure that takes a rep number as input and displays the corresponding rep name.

Figure 8-7 shows a procedure to find the name of the representative whose number is stored in the I_REP_NUM argument. Because the restriction involves the primary key, the query will produce only one row of output. (You will see how to handle queries whose results can contain multiple rows later in this chapter.) The command shown in Figure 8-7 is stored in a script file and is displayed in the Script Editor. To create the procedure, you would run the script file. Assuming that the script file does not contain any errors, Oracle would then create the procedure and it would be available for use.

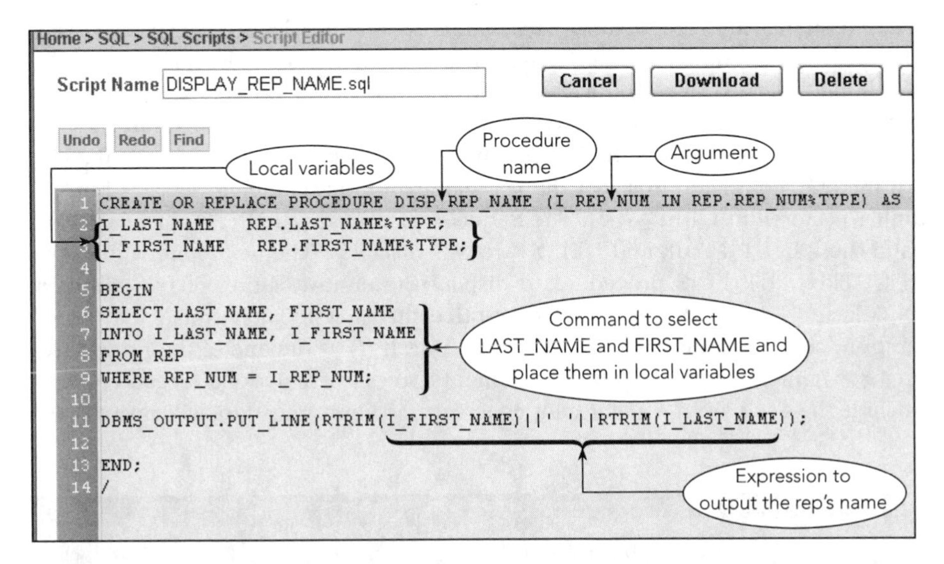

FIGURE 8-7 Procedure to find a rep's name given the rep's number

NOTE

PL/SQL commands, like SQL commands, are free-format and can include blank lines to separate important sections of the procedure and spaces on the lines to make the commands more readable.

The CREATE PROCEDURE command in the stored procedure causes Oracle to create a procedure named DISP_REP_NAME. By including the optional OR REPLACE clause in the CREATE PROCEDURE command, you can use the command to modify an existing procedure. If you omit the OR REPLACE clause, you would need to drop the procedure and then re-create it in order to change the procedure later.

The first line of the command contains a single argument, I_REP_NUM. The word IN following the single argument name indicates that I_REP_NUM will be used for input. That is, the user must enter a value for I_REP_NUM to use the procedure. Other possibilities are OUT, which indicates that the procedure will set a value for the argument, and INOUT, which indicates that the user will enter a value that the procedure can later change.

Variable names in PL/SQL must start with a letter and can contain letters, dollar signs, underscores, and number signs, but cannot exceed 30 characters. When declaring variables, you must assign the variable a data type, just as you do in the SQL CREATE TABLE command. You can ensure that a variable has the same data type as a particular column in a table by using the %TYPE attribute. To do so, you include the name of the table, followed by a period and the name of the column, and then %TYPE. When you use %TYPE, you do not enter a data type because the variable is automatically assigned the same type as the corresponding column. In the first line of the script file shown in Figure 8-7, assigning the variable I_REP_NUM the same type as the REP_NUM column in the REP table is written as REP.REP_NUM%TYPE.

The first line of the CREATE PROCEDURE command ends with the word AS and is followed by the commands in the procedure. The commands on lines 2 and 3 declare the local variables the procedure requires. In Figure 8-7, lines 2 and 3 create two variables named I_LAST_NAME and I_FIRST_NAME. Both variables are assigned data types using %TYPE.

The **procedural code**, which contains the commands that specify the procedure's function, appears between the BEGIN and END commands. In Figure 8-7, the procedural code begins with the SQL command to select the last name and first name of the sales rep whose number is stored in I_REP_NUM. The SQL command uses the INTO clause to place the results in the I_LAST_NAME and I_FIRST_NAME variables. The next command uses the DBMS_OUTPUT.PUT_LINE procedure to display the concatenation of the trimmed I_FIRST_NAME and I_LAST_NAME variables. Notice that a semicolon ends each variable declaration, command, and the word END. The slash (/) at the end of the procedure appears on its own line. In some Oracle environments, the slash is optional. A good practice is to include the slash, even when it's not necessary, so your procedure will always work correctly.

NOTE

DBMS_OUTPUT is a package that contains multiple procedures, including PUT_LINE. The SQL Commands page automatically displays the output produced by DBMS_OUTPUT.

To **call** (or use) the procedure from the SQL Commands page, type the word BEGIN, followed by the name of the procedure including the desired value for the argument in parentheses, followed by the word END, a semicolon, and a slash on a separate line. To use the DISP_REP_NAME procedure to find the name of sales rep 20, for example, type the command shown in Figure 8-8.

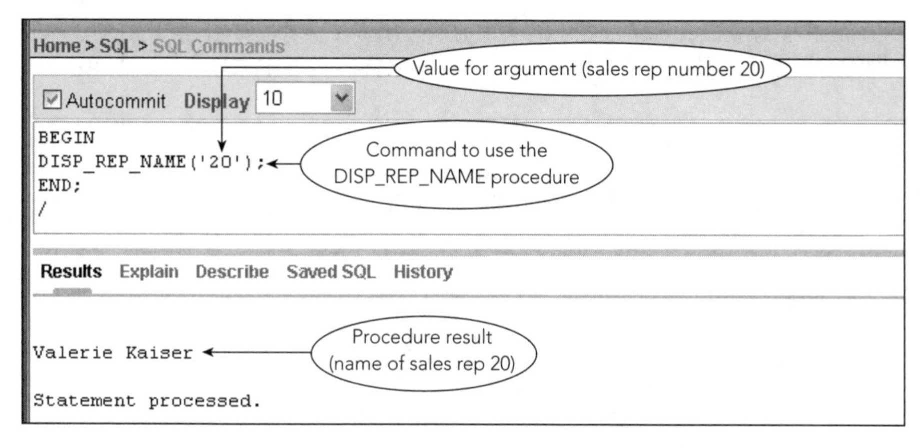

FIGURE 8-8 Using the DISP_REP_NAME procedure within an SQL command

ERROR HANDLING

Procedures must be able to handle conditions that can arise when accessing the database. For example, the user enters a rep number and the DISP_REP_NAME procedure displays the corresponding rep's name. What happens when the user enters an invalid rep number? This situation results in the error message shown in Figure 8-9 because Oracle will not find any last name to display.

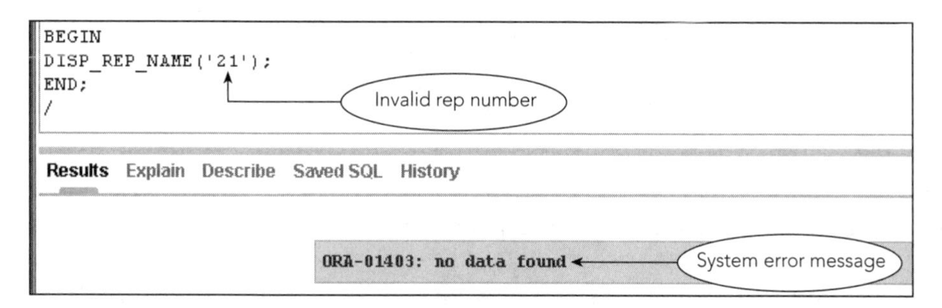

FIGURE 8-9 System error that occurs when a user enters an invalid rep number

You can include the EXCEPTION clause shown in Figure 8-10 to handle processing an invalid rep number. When a user enters a rep number that does not match any rep number in the REP table, the NO_DATA_FOUND condition on line 13 will be true. When the NO_DATA_FOUND condition is true, the procedure displays the "No rep with this number:" message followed by the invalid rep number.

```
CREATE OR REPLACE PROCEDURE DISP_REP_NAME (I_REP_NUM IN REP.REP_NUM%TYPE) AS
I_LAST_NAME    REP.LAST_NAME%TYPE;
I_FIRST_NAME    REP.FIRST_NAME%TYPE;

BEGIN
SELECT LAST_NAME, FIRST_NAME
INTO I_LAST_NAME, I_FIRST_NAME
FROM REP
WHERE REP_NUM = I_REP_NUM;

DBMS_OUTPUT.PUT_LINE(RTRIM(I_FIRST_NAME)||' '||RTRIM(I_LAST_NAME));
EXCEPTION
WHEN NO_DATA_FOUND THEN
DBMS_OUTPUT.PUT_LINE('No rep with this number: '||I_REP_NUM);

END;
/
```

EXCEPTION clause

Tests for NO_DATA_FOUND condition

Action to take when no data is found

FIGURE 8-10 PL/SQL procedure with error handling

When you use this version of the procedure and enter an invalid rep number, you will see the error message from the procedure (Figure 8-11) instead of the system error message (Figure 8-9).

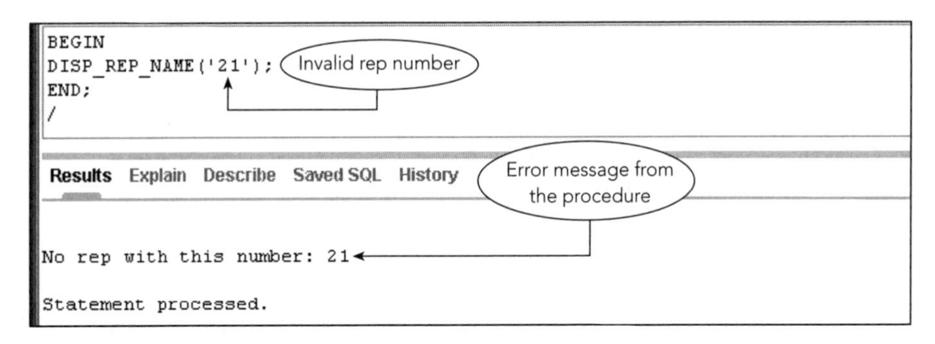

FIGURE 8-11 Error message that occurs when a user enters an invalid rep number

The DISP_REP_NAME procedure must handle an error that results when a user enters an invalid rep number. There are other types of errors that procedures must handle, depending on the processing required. For example, a user might enter a commission rate in a procedure to find the name of the sales rep who has that commission rate. When the user enters the rate 0.05, the procedure will display the TOO_MANY_ROWS error because Valerie Kaiser and Juan Perez both have this same commission rate—the procedure finds two rows instead of one. You can manage this error by writing a WHEN clause that contains a TOO_MANY_ROWS condition, following the EXCEPTION clause in the procedure. You can write both WHEN clauses in the same procedure or in separate procedures. When adding both WHEN clauses to the same procedure, however, the EXCEPTION clause appears only once.

USING UPDATE PROCEDURES

In Chapter 6, you learned how to use SQL commands to update data. You can use the same commands within procedures. A procedure that updates data is called an **update procedure**.

Changing Data with a Procedure

You can use an update procedure to change a row in a table, as illustrated in Example 8.

EXAMPLE 8

Change the name of the customer whose number is stored in I_CUSTOMER_NUM to the value currently stored in I_CUSTOMER_NAME.

This procedure is similar to the procedures used in previous examples with two main differences: it uses an UPDATE command instead of a SELECT command, and there are two arguments, I_CUSTOMER_NUM and I_CUSTOMER_NAME. The I_CUSTOMER_NUM argument stores the customer number to be updated and the I_CUSTOMER_NAME argument stores the new value for the customer name. The procedure appears in Figure 8-12.

```
1  CREATE OR REPLACE PROCEDURE CHG_CUST_NAME (I_CUSTOMER_NUM IN CUSTOMER.CUSTOMER_NUM%TYPE,
2  I_CUSTOMER_NAME IN CUSTOMER.CUSTOMER_NAME%TYPE) AS
3
4  BEGIN
5  UPDATE CUSTOMER
6  SET CUSTOMER_NAME = I_CUSTOMER_NAME
7  WHERE CUSTOMER_NUM = I_CUSTOMER_NUM;
8
9  END;
10 /
```

Arguments

FIGURE 8-12 Using a procedure to update a row

When you run this procedure, you will need to furnish values for two arguments. Figure 8-13 uses this procedure to change the name of customer 725 to Deerfield's.

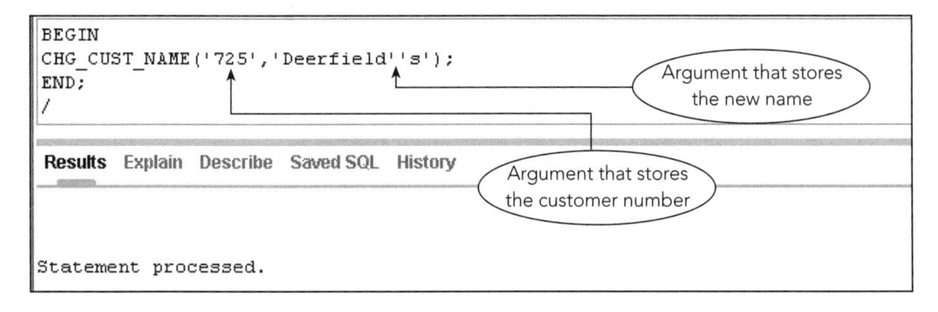

FIGURE 8-13 Using a procedure to update the name of customer 725

Deleting Data with a Procedure

Just as you would expect, if you can use an update procedure to change a rows in a table, you can also use one to delete a row from a table, as illustrated in Example 9.

EXAMPLE 9

Delete the order whose number is stored in I_ORDER_NUM from the ORDERS table, and also delete each order line for the order whose order number is currently stored in the variable from the ORDER_LINE table.

If you attempt to delete the order in the ORDERS table first, referential integrity will prevent the deletion because matching rows would still exist in the ORDER_LINE table, so it is a good idea to delete the orders from the ORDER_LINE table first. The procedure to delete an order and its related order lines appears in Figure 8-14. This procedure contains two DELETE commands. The first command deletes all order lines in the ORDER_LINE table on which the order number matches the value stored in the I_ORDER_NUM argument. The second command deletes the order in the ORDERS table whose order number matches the value stored in the I_ORDER_NUM argument.

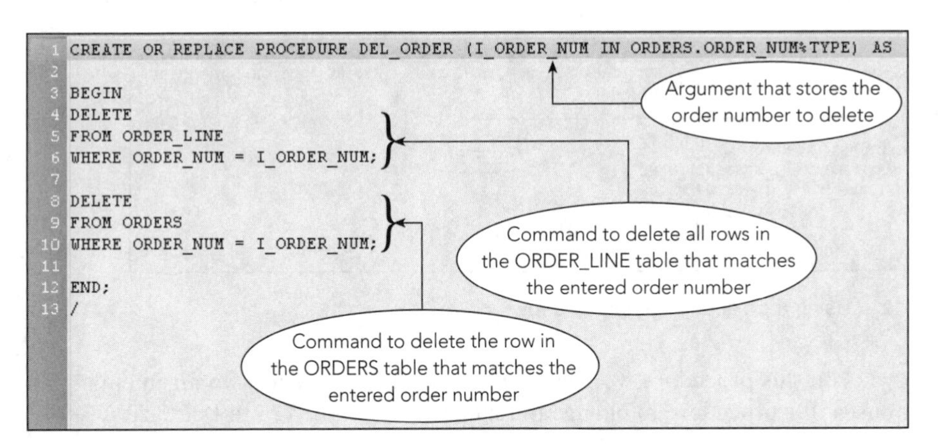

FIGURE 8-14 Procedure to delete a row and related rows from multiple tables

Figure 8-15 shows the use of this procedure to delete order number 21610. Even though there are two DELETE commands in the procedure, the user enters the order number only once.

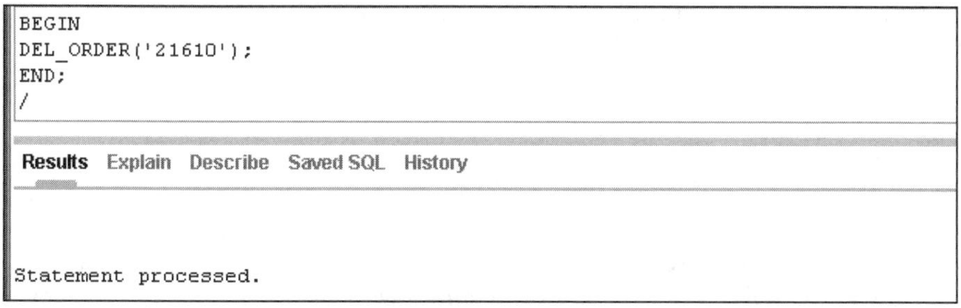

```
BEGIN
DEL_ORDER('21610');
END;
/
```

Results Explain Describe Saved SQL History

Statement processed.

FIGURE 8-15 Using the procedure to delete an order

SELECTING MULTIPLE ROWS WITH A PROCEDURE

The procedures you have seen so far include commands that retrieve individual rows. You can use an UPDATE or a DELETE command in PL/SQL to update or delete multiple rows. The commands are executed and the updates or deletions occur. Then the procedure can move on to the next task.

What happens when a SELECT command in a procedure retrieves multiple rows? For example, suppose the SELECT command retrieves the number and name of each customer represented by the sales rep whose number is stored in I_REP_NUM. There is a problem—PL/SQL can process only one record at a time, but this SQL command retrieves more than one row. Whose number and name is placed in I_CUSTOMER_NUM and I_CUSTOMER_NAME when the command retrieves more than one customer row? Should you make I_CUSTOMER_NUM and I_CUSTOMER_NAME arrays capable of holding multiple rows and, if so, what should be the size of these arrays? Fortunately, you can solve this problem by using a cursor.

Using a Cursor

A **cursor** is a pointer to a row in the collection of rows retrieved by an SQL command. (This is *not* the same cursor that you see on your computer screen.) The cursor advances one row at a time to provide sequential, one-record-at-a-time access to the retrieved rows so PL/SQL can process the rows. By using a cursor, PL/SQL can process the set of retrieved rows as though they were records in a sequential file.

To use a cursor, you must first declare it, as illustrated in Example 10.

EXAMPLE 10

Retrieve and list the number and name of each customer represented by the sales rep whose number is stored in the variable I_REP_NUM.

The first step in using a cursor is to declare the cursor and describe the associated query in the declaration section of the procedure. In this example, assuming the cursor is named CUSTGROUP, the command to declare the cursor is:

```
CURSOR CUSTGROUP IS
SELECT CUSTOMER_NUM, CUSTOMER_NAME
FROM CUSTOMER
WHERE REP_NUM = I_REP_NUM;
```

This command does *not* cause the query to be executed at this time; it only declares a cursor named CUSTGROUP and associates the cursor with the indicated query. Using a cursor in a procedure involves three commands: OPEN, FETCH, and CLOSE. The **OPEN** command opens the cursor and causes the query to be executed, making the results available to the procedure. Executing a **FETCH** command advances the cursor to the next row in the set of rows retrieved by the query and places the contents of the row in the indicated variables. Finally, the **CLOSE** command closes a cursor and deactivates it. Data retrieved by the execution of the query is no longer available. The cursor could be opened again later and processing could begin again.

The OPEN, FETCH, and CLOSE commands used in processing a cursor are analogous to the OPEN, READ, and CLOSE commands used in processing a sequential file.

Opening a Cursor

Prior to opening the cursor, there are no rows available to be fetched. In Figure 8-16, this is indicated by the absence of data in the CUSTGROUP portion of the figure. The right side of the figure illustrates the variables into which the data will be placed (I_CUSTOMER_NUM and I_CUSTOMER_NAME) and the value CUSTGROUP%NOTFOUND. Once the cursor has been opened and all the records have been fetched, the CUSTGROUP%NOTFOUND value is set to TRUE. Procedures using the cursor can use this value to indicate when the fetching of rows is complete.

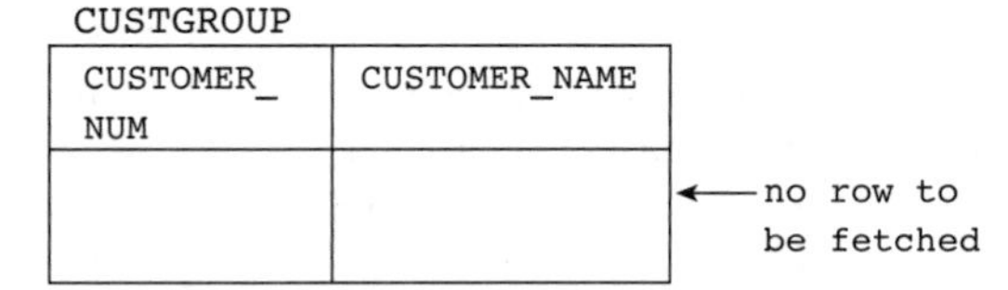

I_CUSTOMER_NUM	I_CUSTOMER_NAME	CUSTGROUP %NOTFOUND
		FALSE

FIGURE 8-16 Before OPEN

The OPEN command is written as follows:

```
OPEN CUSTGROUP;
```

Figure 8-17 shows the result of opening the CUSTGROUP cursor. In the figure, assume that I_REP_NUM is set to 20 before the OPEN command is executed; there are now three rows available to be fetched. No rows have yet been fetched, as indicated by the absence of values in I_CUSTOMER_NUM and I_CUSTOMER_NAME. CUSTGROUP%NOTFOUND is still FALSE. The cursor is positioned at the first row; that is, the next FETCH command causes the contents of the first row to be placed in the indicated variables.

CUSTGROUP

CUSTOMER_ NUM	CUSTOMER_NAME	
148	Al's Appliance and Sport	← next row to be fetched
524	Kline's	
842	All Season	

I_CUSTOMER_ NUM	I_CUSTOMER_NAME	CUSTGROUP %NOTFOUND
		FALSE

FIGURE 8-17 After OPEN, but before first FETCH

Fetching Rows from a Cursor

To fetch (get) the next row from a cursor, use the FETCH command. The FETCH command is written as follows:

```
FETCH CUSTGROUP INTO I_CUSTOMER_NUM, I_CUSTOMER_NAME;
```

Note that the INTO clause is associated with the FETCH command itself and not with the query used in the cursor definition. The execution of this query could produce multiple rows. The execution of the FETCH command produces only a single row, so it is appropriate that the FETCH command causes data to be placed in the indicated variables.

Figure 8-18 through Figure 8-21 show the result of four FETCH commands. The first three fetches are successful. In each case, the data from the appropriate row in the cursor is placed in the indicated variables and CUSTGROUP%NOTFOUND is still FALSE. The fourth FETCH command is different, however, because there is no more data to fetch. In this case, the contents of the variables are left untouched and CUSTGROUP%NOTFOUND is set to TRUE.

CUSTGROUP

CUSTOMER_ NUM	CUSTOMER_NAME	
148	Al's Appliance and Sport	
524	Kline's	← next row to be fetched
842	All Season	

I_CUSTOMER_ NUM	I_CUSTOMER_NAME	CUSTGROUP %NOTFOUND
148	Al's Appliance and Sport	FALSE

FIGURE 8-18 After first FETCH

CUSTGROUP

CUSTOMER_ NUM	CUSTOMER_NAME	
148	Al's Appliance and Sport	
524	Kline's	
842	All Season	← next row to be fetched

I_CUSTOMER_ NUM	I_CUSTOMER_NAME	CUSTGROUP %NOTFOUND
524	Kline's	FALSE

FIGURE 8-19 After second FETCH

CUSTGROUP

CUSTOMER_NUM	CUSTOMER_NAME
148	Al's Appliance and Sport
524	Kline's
842	All Season

← next row to be fetched

I_CUSTOMER_NUM	I_CUSTOMER_NAME	CUSTGROUP %NOTFOUND
842	All Season	FALSE

FIGURE 8-20 After third FETCH

CUSTGROUP

CUSTOMER_NUM	CUSTOMER_NAME
148	Al's Appliance and Sport
524	Kline's
842	All Season

← no more rows to be fetched

I_CUSTOMER_NUM	I_CUSTOMER_NAME	CUSTGROUP %NOTFOUND
842	All Season	TRUE

FIGURE 8-21 After attempting a fourth FETCH (CUSTGROUP%NOTFOUND is TRUE)

Closing a Cursor

The CLOSE command is written as follows:

```
CLOSE CUSTGROUP;
```

Figure 8-22 shows the result of closing the CUSTGROUP cursor. The data is no longer available.

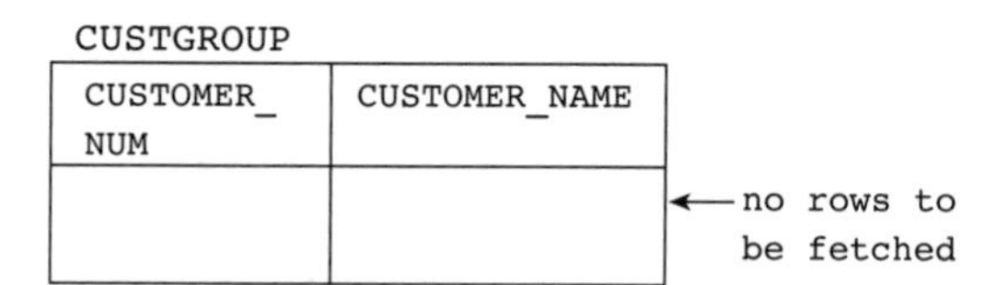

FIGURE 8-22 After CLOSE

Writing a Complete Procedure Using a Cursor

Figure 8-23 shows a complete procedure using a cursor. The declaration portion contains the CUSTGROUP cursor definition. The procedural portion begins with the command to open the CUSTGROUP cursor. The statements between the LOOP and END LOOP commands create a loop that begins by fetching the next row from the cursor and placing the results in I_CUSTOMER_NUM and I_CUSTOMER_NAME. The EXIT command tests the condition CUSTGROUP%NOTFOUND. If the condition is true, the loop is terminated. If the condition is not true, the DBMS_OUTPUT.PUT_LINE commands display the contents of I_CUSTOMER_NUM and I_CUSTOMER_NAME.

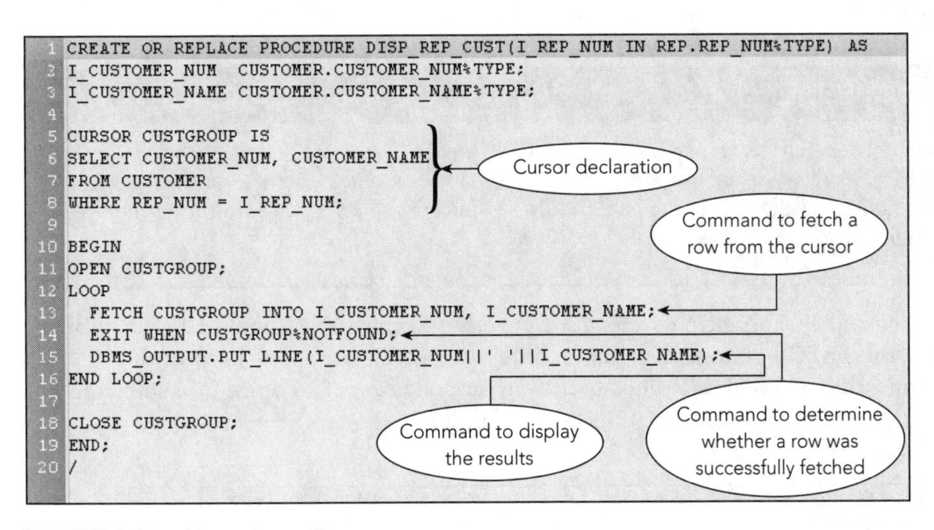

FIGURE 8-23 Procedure with a cursor

Figure 8-24 shows the results of using the procedure. After the user enters 20 as the value for the rep number, the procedure displays the number and name of each customer of sales rep 20.

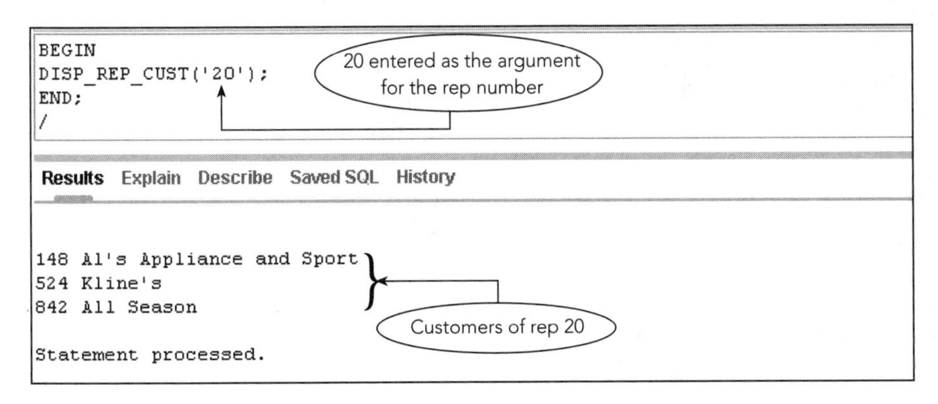

FIGURE 8-24 Results of using the procedure

Using More Complex Cursors

The query formulation that defined the cursor in Example 10 was straightforward. Any SQL query is legitimate in a cursor definition. In fact, the more complicated the requirements for retrieval, the more numerous the benefits derived by the programmer who uses embedded SQL. Consider the query in Example 11.

EXAMPLE 11

For each order that contains an order line for the part whose part number is stored in I_PART_NUM, retrieve the order number, order date, customer number, name of the customer that placed the order, and last and first names of the sales rep who represents the customer.

Opening and closing the cursor is done exactly as shown in Example 10. The only difference in the FETCH command is that a different set of variables is used in the INTO clause. Thus, the only real difference is the cursor definition. The procedure shown in Figure 8-25 contains the appropriate cursor definition.

```
1  CREATE OR REPLACE PROCEDURE DISP_PART_ORDERS (I_PART_NUM IN PART.PART_NUM%TYPE) AS
2
3  I_ORDER_NUM ORDERS.ORDER_NUM%TYPE;
4  I_ORDER_DATE ORDERS.ORDER_DATE%TYPE;
5  I_CUSTOMER_NUM ORDERS.CUSTOMER_NUM%TYPE;
6  I_REP_NUM REP.REP_NUM%TYPE;
7  I_LAST_NAME REP.LAST_NAME%TYPE;
8  I_FIRST_NAME REP.FIRST_NAME%TYPE;
9
10 CURSOR ORDGROUP IS
11 SELECT ORDERS.ORDER_NUM, ORDER_DATE, ORDERS.CUSTOMER_NUM, CUSTOMER.REP_NUM,
12 LAST_NAME, FIRST_NAME
13 FROM ORDER_LINE, ORDERS, CUSTOMER, REP
14 WHERE ORDER_LINE.ORDER_NUM = ORDERS.ORDER_NUM
15 AND ORDERS.CUSTOMER_NUM = CUSTOMER.CUSTOMER_NUM
16 AND CUSTOMER.REP_NUM = REP.REP_NUM
17 AND PART_NUM = I_PART_NUM;
18
19 BEGIN
20
21 OPEN ORDGROUP;
22 LOOP
23   FETCH ORDGROUP INTO I_ORDER_NUM, I_ORDER_DATE, I_CUSTOMER_NUM, I_REP_NUM,
24   I_LAST_NAME, I_FIRST_NAME;
25   EXIT WHEN ORDGROUP%NOTFOUND;
26
27   DBMS_OUTPUT.PUT_LINE(I_ORDER_NUM);
28   DBMS_OUTPUT.PUT_LINE(I_ORDER_DATE);
29   DBMS_OUTPUT.PUT_LINE(I_CUSTOMER_NUM);
30   DBMS_OUTPUT.PUT_LINE(I_LAST_NAME);
31   DBMS_OUTPUT.PUT_LINE(I_FIRST_NAME);
32
33 END LOOP;
34
35 CLOSE ORDGROUP;
36 END;
37 /
```

FIGURE 8-25 Procedure with a cursor that involves joining multiple tables

The results of using this procedure to display the results for part DR93 are shown in Figure 8-26.

```
BEGIN
DISP_PART_ORDERS('DR93');
END;
/
```

Results Explain Describe Saved SQL History

```
21610
20-OCT-10
356
Perez
Juan
21619
23-OCT-10
148
Kaiser
Valerie

Statement processed.
```

FIGURE 8-26 Results of using the procedure to display orders containing part DR93

Advantages of Cursors

The retrieval requirements in Example 11 are substantial. Beyond coding the preceding cursor definition, the programmer doesn't need to worry about the mechanics of obtaining the necessary data or placing it in the right order, because this happens automatically when the cursor is opened. To the programmer, it seems as if a sequential file already exists that contains the correct data, sorted in the right order. This assumption leads to three main advantages:

1. The coding in the procedure is greatly simplified.
2. In a normal program, the programmer must determine the most efficient way to access the data. In a program or procedure using embedded SQL, the optimizer determines the best way to access the data. The programmer isn't concerned with the best way to retrieve the data. In addition, when an underlying structure changes (for example, an additional index is created), the optimizer determines the best way to execute the query with the new structure. The program or procedure does not have to change at all.
3. When the database structure changes in such a way that the necessary information is still obtainable using a different query, the only change required in the program or procedure is the cursor definition. The procedural code is not affected.

USING T-SQL IN SQL SERVER

SQL Server uses an extended version of SQL called T-SQL (Transact-SQL). You can use T-SQL to create stored procedures and use cursors. The reasons for creating and using stored procedures and cursors are identical to those discussed in the PL/SQL section. Only the command syntax is different.

Retrieving a Single Row and Column

In Example 7, you learned how to write a procedure in PL/SQL that takes a rep number as input and displays the corresponding rep name. The following code shows how you would create the stored procedure in T-SQL:


```
CREATE PROCEDURE usp_DISP_REP_NAME
@repnum char(2)
AS
SELECT RTRIM(FIRST_NAME)+' '+RTRIM(LAST_NAME)
FROM REP
WHERE REP_NUM = @repnum
```

The CREATE PROCEDURE command in the stored procedure causes SQL Server to create a procedure named usp_DISP_REP_NAME. The usp_ prefix identifies the procedure as a user-stored procedure. Although using the prefix is optional, it is an easy way to differentiate user-stored procedures from SQL Server system-stored procedures. The argument for this procedure is @repnum. In T-SQL, you must assign a data type to parameters. All arguments start with the at (@) sign. Arguments should have the same data type and length as the particular column in a table that they represent. In the REP table, REP_NUM was defined with a CHAR data type and a length of 2. The CREATE PROCEDURE ends with the word AS followed by the SELECT command that comprises the procedure.

To call the procedure, use the EXEC command and include any arguments in single quotes. The procedure to find the name of sales rep 20 is:

```
EXEC usp_DISP_REP_NAME'20'
```

The result of executing this procedure is the same as that shown in Figure 8-8.

Changing Data with a Stored Procedure

In Example 8, you learned how to write a procedure in PL/SQL that changes the name of a customer. The following commands show how to create the stored procedure in T-SQL:

```
CREATE PROCEDURE usp_CHG_CUST_NAME
@custnum char(3),
@custname char(35)
AS
UPDATE CUSTOMER
SET CUSTOMER_NAME = @custname
WHERE CUSTOMER_NUM = @custnum
```

The procedure has two arguments, @custnum and @custname, and uses an UPDATE command instead of a SELECT command. To execute a stored procedure with two arguments, separate the arguments with a comma as shown in the following command:

```
EXEC usp_CHG_CUST_NAME'725','Deerfield''s'
```

Deleting Data with a Stored Procedure

In Example 9, you learned how to write a procedure in PL/SQL that deletes an order number from both the ORDER_LINE table and the ORDERS table. The following commands show how to create the stored procedure in T-SQL:

```
CREATE PROCEDURE usp_DEL_ORDER
@ordernum char(5)
AS
DELETE
FROM
ORDER_LINE
WHERE ORDER_NUM = @ordernum

DELETE
FROM ORDERS
WHERE ORDER_NUM = @ordernum
```


Using a Cursor

Cursors serve the same purpose in T-SQL as they do in PL/SQL and work exactly the same way. You need to declare a cursor, open a cursor, fetch rows from a cursor, and close a cursor. The only difference is in the command syntax. The following T-SQL code performs exactly the same task as that shown in Example 10:

```
CREATE PROCEDURE usp_DISP_REP_CUST
@repnum char(2)
AS
DECLARE @custnum char(3)
DECLARE @custname char(35)
DECLARE mycursor CURSOR READ_ONLY

FOR
SELECT CUSTOMER_NUM, CUSTOMER_NAME
FROM CUSTOMER
WHERE REP_NUM = @repnum

OPEN mycursor

FETCH NEXT FROM mycursor
INTO @custnum, @custname

WHILE @@FETCH_STATUS = 0
BEGIN

     PRINT @custnum+' '+@custname

     FETCH NEXT FROM mycursor
     INTO @custnum, @custname
END

CLOSE mycursor
DEALLOCATE mycursor
```

The procedure uses one argument, @repnum. It also uses two variables, and each variable must be declared using a DECLARE statement. You also declare the cursor by giving it a name, describing its properties, and associating it with a SELECT statement. The cursor property, READ_ONLY, means that the cursor is used for retrieval purposes only. The OPEN, FETCH, and CLOSE commands perform exactly the same tasks in T-SQL as they do in PL/SQL. The OPEN command opens the cursor and causes the query to be executed. The FETCH command advances the cursor to the next row and places the contents of the row in the indicated variables. The CLOSE command closes a cursor and the DEALLOCATE command deletes the cursor. The DEALLOCATE command is not necessary but it does enable the user to use the same cursor name with another procedure.

The WHILE loop will repeat until the value of the system variable @@FETCH_STATUS is not zero. The PRINT command will output the values stored in the @custnum and @custname variables.

Using More Complex Cursors

T-SQL also can handle more complex queries. The T-SQL code for Example 11 is shown below:

```
CREATE PROCEDURE usp_DISP_PART_ORDERS
@partnum char(4)
AS
DECLARE @ordernum char(5)
DECLARE @orderdate datetime
DECLARE @custnum char(3)
DECLARE @repnum char(2)
DECLARE @lastname char(15)
DECLARE @firstname char(15)

DECLARE mycursor CURSOR READ_ONLY

FOR
SELECT ORDERS.ORDER_NUM, ORDER_DATE, ORDERS.CUSTOMER_NUM,
     CUSTOMER.REP_NUM, LAST_NAME, FIRST_NAME
FROM ORDER_LINE, ORDERS, CUSTOMER, REP
WHERE ORDER_LINE.ORDER_NUM = ORDERS.ORDER_NUM
AND ORDERS.CUSTOMER_NUM = CUSTOMER.CUSTOMER_NUM
AND CUSTOMER.REP_NUM = REP.REP_NUM
AND PART_NUM = @partnum

OPEN mycursor

FETCH NEXT FROM mycursor
INTO @ordernum, @orderdate, @custnum, @repnum, @lastname, @firstname

WHILE @@FETCH_STATUS = 0
BEGIN

     PRINT @ordernum
     PRINT @orderdate
     PRINT @custnum
     PRINT @lastname
     PRINT @firstname

     FETCH NEXT FROM mycursor
     INTO @ordernum, @orderdate, @custnum, @repnum, @lastname, @firstname
END

CLOSE mycursor
DEALLOCATE mycursor
```

USING SQL IN MICROSOFT ACCESS

Not every programming language accepts SQL commands as readily as PL/SQL and T-SQL. In Microsoft Access, programs are written in Visual Basic, which does not support embedded SQL commands directly in the code. When the SQL command is stored in a string variable, however, you can use the DoCmd.RunSQL command to run the command. The procedure in which you place the SQL command can include arguments.

Deleting Data with Visual Basic

To delete the sales rep whose number is 20, the command is:

```
DELETE FROM REP WHERE REP_NUM = '20';
```


When you write this type of command, you usually don't know in advance the specific sales rep number that you want to delete; it would be passed as an argument to the procedure containing this DELETE command. In the following example, the sales rep number is stored in an argument named I_REP_NUM.

EXAMPLE 12

Delete from the REP table the sales rep whose number currently is stored in I_REP_NUM.

Statements in the procedure usually create the appropriate DELETE command, using the value in any necessary arguments. For example, when the command is stored in the variable named strSQL (which must be a string variable) and the rep number is stored in the argument I_REP_NUM, the following command is appropriate:

```
strSQL = "DELETE FROM REP WHERE REP_NUM = '"
strSQL = strSQL & I_REP_NUM
strSQL = strSQL & "';"
```

The first command sets the strSQL string variable to DELETE FROM REP WHERE REP_NUM = '; that is, it creates everything necessary in the command up to and including the single quotation mark preceding the rep number. The second command uses concatenation (&). It changes strSQL to the result of the previous value concatenated with the value in I_REP_NUM. When I_REP_NUM contains the value 20, for example, the command would be DELETE FROM REP WHERE REP_NUM = '20. The final command sets strSQL to the result of the value already created, concatenated with a single quotation mark and a semicolon. The command is now complete.

Figure 8-27 shows a completed procedure to accomplish the necessary deletion in Access. You enter this procedure in the Microsoft Visual Basic window. In the program, the Dim statement creates a string variable named strSQL. The next three commands set strSQL to the appropriate SQL command. Finally, the DoCmd.RunSQL command runs the SQL command stored in strSQL.

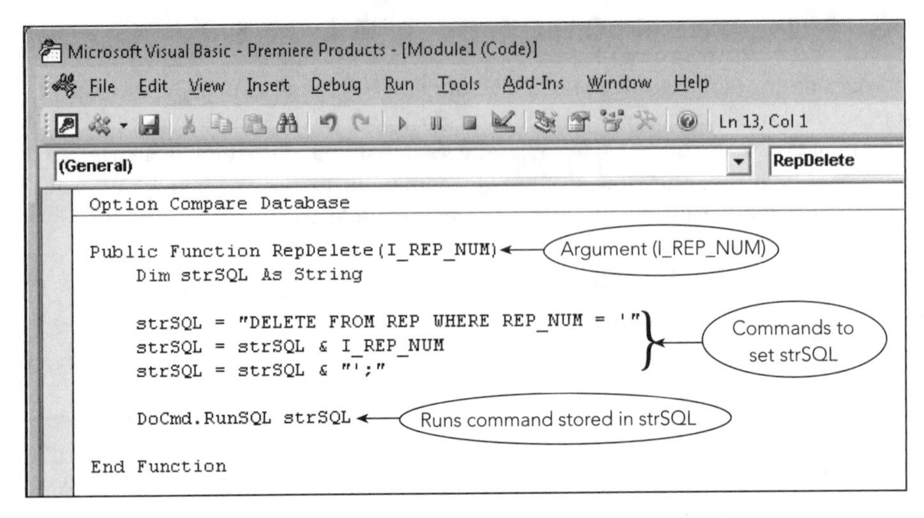

FIGURE 8-27 Visual Basic code to delete a sales rep

NOTE

If you have concerns about how you constructed the SQL command in strSQL, you can include the Debug.Print (strSQL) command after the set of commands that construct strSQL. The Debug.Print command displays the entire command before it is executed so you can review it for accuracy. If you need to correct an error, rerun the program after making the necessary changes. If you get an error in your program, check your SQL command carefully to make sure that you concatenated it correctly.

Running the Code

Normally, you run code like the function shown in Figure 8-27 by calling it from another procedure or associating it with some event, such as clicking a button on a form. However, you can run it directly by using the Immediate window (click View on the menu bar, and then click Immediate Window to open it). Normally, you would use this window only for testing purposes, but you can use it to see the result of running the code. To run a Function procedure, such as the one shown in Figure 8-27, in the Immediate window, type a question mark followed by the name of the procedure and a set of parentheses, as shown in Figure 8-28. Place the values for any arguments in the parentheses. Assuming that you wanted to delete a sales rep whose number is 50, you would include "50" inside the parentheses as shown in the figure.

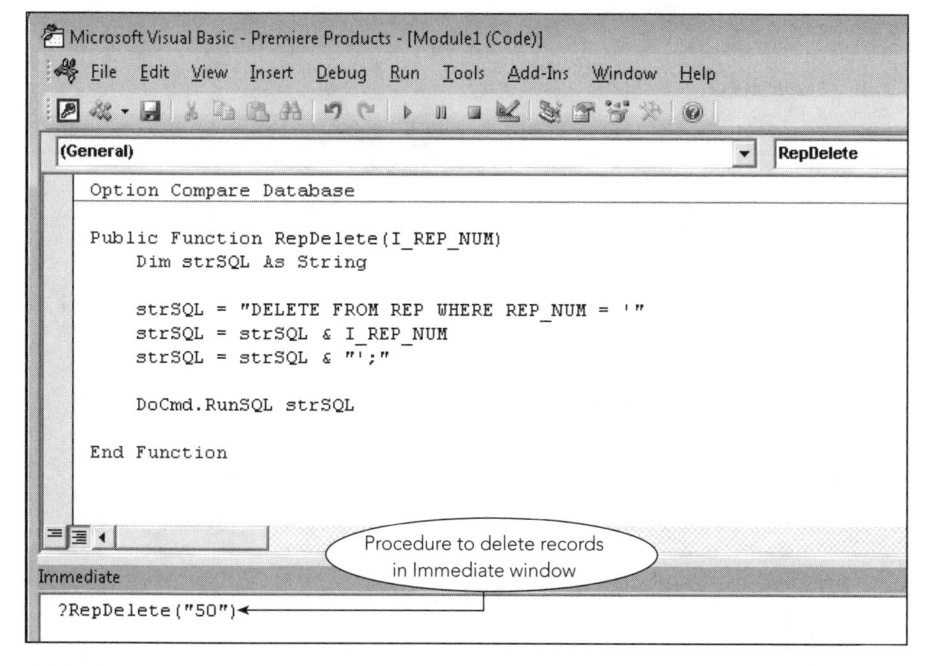

FIGURE 8-28 Running the code in the Immediate window

After you type the command and press the Enter key, the code will run and the appropriate action will occur. In this case, the command deletes the sales rep with the number 50 (assuming there is a sales rep 50).

Updating Data with Visual Basic

A procedure that updates a table using an UPDATE command is similar to the one used to delete a sales rep. In Example 13, two arguments are required. One of them, I_LAST_NAME, contains the new name for the sales rep. The other, I_REP_NUM, contains the number of the rep whose name is to be changed.

EXAMPLE 13

Change the last name of the sales rep whose number is stored in I_REP_NUM to the value currently stored in I_LAST_NAME.

This example is similar to the previous one with two important differences. First, you need to use the UPDATE command instead of the DELETE command. Second, there are two arguments, so there are two portions of the construction of the SQL command that involve variables. The complete procedure is shown in Figure 8-29.

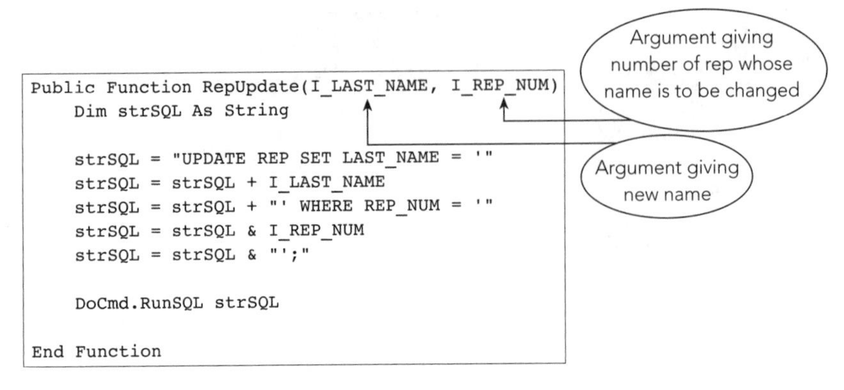


```
Public Function RepUpdate(I_LAST_NAME, I_REP_NUM)
    Dim strSQL As String

    strSQL = "UPDATE REP SET LAST_NAME = '"
    strSQL = strSQL + I_LAST_NAME
    strSQL = strSQL + "' WHERE REP_NUM = '"
    strSQL = strSQL & I_REP_NUM
    strSQL = strSQL & "';"

    DoCmd.RunSQL strSQL

End Function
```


FIGURE 8-29 Code to change a rep's last name

To run this procedure, you would enter values for both arguments as shown in Figure 8-30.

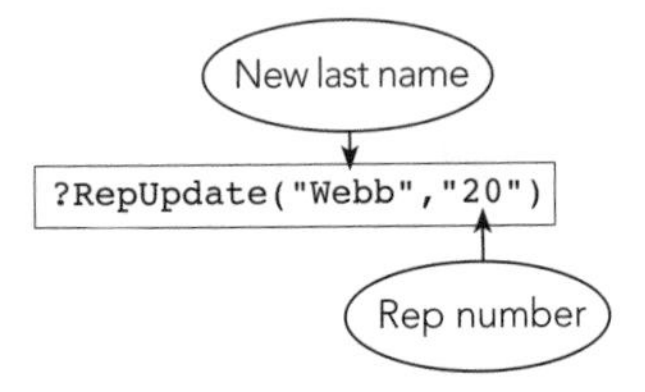


```
?RepUpdate("Webb","20")
```

FIGURE 8-30 Running the code to change a rep's last name

Inserting Data with Visual Basic

The process for inserting rows is similar in Access when compared to PL/SQL or T-SQL. You create the appropriate INSERT command in the strSQL variable. There will be multiple arguments in the procedure—one for each value to be inserted.

Finding Multiple Rows with Visual Basic

Just as when embedding SQL in PL/SQL, deleting or updating multiple rows causes no problems, because these procedures still represent a single operation, with all the work happening behind the scenes. A SELECT command that returns several rows, however, poses serious problems for record-at-a-time languages like PL/SQL and Visual Basic. You handle SELECT commands differently in Access than you do in PL/SQL or T-SQL. In particular, there are no cursors in Access. Instead, you handle the results of a query just as you might use a loop to process through the records in a table.

EXAMPLE 14

Retrieve and list the number and name of each customer represented by the sales rep whose number is stored in the variable I_REP_NUM.

Figure 8-31 shows a procedure to accomplish the indicated task. The statements involving rs and cnn are a typical way of processing through a recordset, that is, through all the records contained in a table or in the results of a query. The only difference between this program and one to process all the records in a table is that the Open command refers to an SQL command and not a table. (The SQL command is stored in the variable named strSQL and is created in the same manner as shown in the previous examples.)

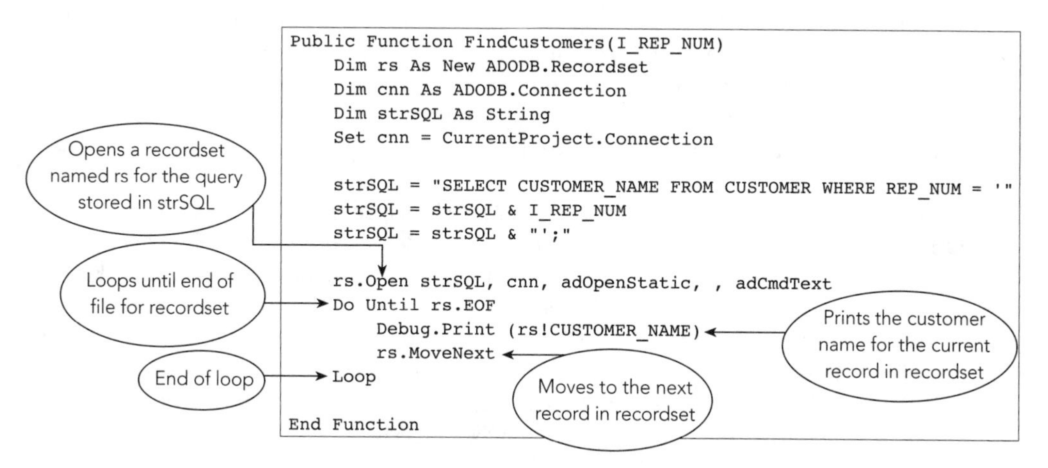

FIGURE 8-31 Code to find customers of a specific rep

The loop continues until reaching the end of file for the recordset, that is, until all records have been processed. Within the loop, you can use the Debug.Print command to print a value. In this case, the value to be printed is rs!CUSTOMER_NAME. This indicates the contents of the CUSTOMER_NAME column for the record in the recordset (rs) on which Access is currently positioned. The next command, rs.MoveNext, moves to the next record in the recordset. The loop continues until all records in the recordset have been processed.

Figure 8-32 shows the results of running this procedure and entering a value of "35" as an argument. Access displays the four customers of sales rep 35.

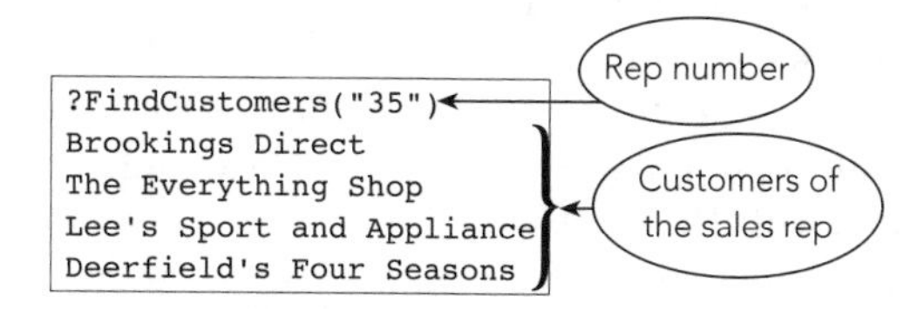

FIGURE 8-32 Running the code to find customers of a sales rep

NOTE

When you expect an SQL query to return only one record, you use the same process but would not need a loop.

USING A TRIGGER

A **trigger** is a procedure that is executed automatically in response to an associated database operation, such as an INSERT, UPDATE, or DELETE command. Unlike a stored procedure, which is executed in response to a user request, a trigger is executed in response to a command that causes the associated database operation to occur.

The examples in this section assume there is a new column named ON_ORDER in the PART table. This column represents the number of units of a part currently on order. For example, if there are two separate order lines for a part and the number ordered on one order line is 3 and the number ordered on the other order line is 2, the ON_ORDER value for that part will be 5. Adding, changing, or deleting order lines affects the value in the ON_ORDER column for the part. To ensure that the value is updated appropriately, you can use a trigger.

If you created the ADD_ORDER_LINE trigger shown in Figure 8-33, the SQL command in the trigger would be executed when a user adds an order line. The trigger must update the ON_ORDER value for the corresponding part to reflect the order line. For example, if the value in the ON_ORDER column for part CD52 is 4 and the user adds an order line on which the part number is CD52 and the number of units ordered is 2, six units of part CD52 will be on order. When a record is added to the ORDER_LINE table, the ADD_ORDER_LINE trigger updates the PART table by adding the number of units ordered on the order line to the previous value in the ON_ORDER column.

```
1 CREATE OR REPLACE TRIGGER ADD_ORDER_LINE
2 AFTER INSERT ON ORDER_LINE FOR EACH ROW
3 BEGIN
4 UPDATE PART
5 SET ON_ORDER = ON_ORDER + :NEW.NUM_ORDERED ;
6 END;
```

FIGURE 8-33 ADD_ORDER_LINE trigger

The first line indicates that the command is creating a trigger named ADD_ORDER_LINE. The second line indicates that this trigger will be executed after an order line is inserted and that the SQL command is to occur for each row that is added. Like stored procedures, the SQL command is enclosed between the words BEGIN and END. In this case, the SQL command is an UPDATE command. The command uses the NEW qualifier, which refers to the row that is added to the ORDER_LINE table. If an order line is added on which the part number is CD52 and the number ordered is 2, for example, NEW.PART_NUM will be CD52 and NEW.NUM_ORDERED will be 2.

The following UPDATE_ORDER_LINE trigger shown in Figure 8-34 is executed when a user attempts to update an order line. There are two differences between the UPDATE_ORDER_LINE trigger and the ADD_ORDER_LINE trigger. First, the second line of the UPDATE_ORDER_LINE trigger indicates that this trigger is executed after an UPDATE of an order line rather than an INSERT. Second, the computation to update the ON_ORDER column includes both NEW.NUM_ORDERED and OLD.NUM_ORDERED. As with the ADD_ORDER_LINE trigger, NEW.NUM_ORDERED refers to the new value. In an

UPDATE command, however, there is also an old value, which is the value before the update takes place. If an update changes the value for ON_ORDER from 1 to 3, OLD.NUM_ORDERED is 1 and NEW.NUM_ORDERED is 3. Adding NEW.NUM_ORDERED and subtracting OLD.NUM_ORDERED results in a net change of an increase of 2. (The net change could also be negative, in which case the ON_ORDER value decreases.)

```
1 CREATE OR REPLACE TRIGGER UDPATE_ORDER_LINE
2 AFTER UPDATE ON ORDER_LINE FOR EACH ROW
3 BEGIN
4 UPDATE PART
5 SET ON_ORDER = ON_ORDER + :NEW.NUM_ORDERED - :OLD.NUM_ORDERED;
6 END;
```

FIGURE 8-34 UPDATE_ORDER_LINE trigger

The DELETE_ORDER_LINE trigger shown in Figure 8-35 performs a function similar to the other two. When an order line is deleted, the ON_ORDER value for the corresponding part is updated by subtracting OLD.NUM_ORDERED from the current ON_ORDER value. (In a delete operation, there is no NEW.NUM_ORDERED.)

```
1 CREATE OR REPLACE TRIGGER DELETE_ORDER_LINE
2 AFTER DELETE ON ORDER_LINE FOR EACH ROW
3 BEGIN
4 UPDATE PART
5 SET ON_ORDER = ON_ORDER - :OLD.NUM_ORDERED ;
6 END;
```

FIGURE 8-35 DELETE_ORDER_LINE trigger

ACCESS USER NOTE

Access does not support triggers. When using a form to update table data, you can achieve some of the same functionality by creating VBA code to be executed after the insertion, update, or deletion of records.

SQL SERVER USER NOTE

In SQL Server, you create triggers using T-SQL. The code to create the ADD_ORDER_LINE trigger is:

```
CREATE TRIGGER ADD_ORDER_LINE
ON ORDER_LINE
AFTER INSERT
AS

DECLARE @numbord decimal(3,0)
SELECT @numbord = (SELECT NUM_ORDERED FROM INSERTED)
UPDATE PART
SET ON_ORDER = ON_ORDER + @numbord
```

This trigger uses one variable, @numbord, and the value placed in that variable is obtained from the SELECT statement. The INSERTED table is a temporary system table that contains a copy of the values that the last SQL command inserted. The column names are the same column names as in the ORDER_LINE table. The INSERTED table holds the most recent value of the NUM_ORDERED column which is what you need to update the PART table.

The T-SQL trigger that executes after an UPDATE of an order line is:

```
CREATE TRIGGER UPDATE_ORDER_LINE
ON ORDER_LINE
AFTER UPDATE
AS

DECLARE @newnumbord decimal(3,0)
DECLARE @oldnumbord decimal(3,0)
SELECT @newnumbord = (SELECT NUM_ORDERED FROM INSERTED)
SELECT @oldnumbord = (SELECT NUM_ORDERED FROM DELETED)
UPDATE PART
SET ON_ORDER = ON_ORDER + @newnumbord - @oldnumbord
```

This trigger uses the INSERTED table and the DELETED table. The DELETED table contains the previous value of the NUM_ORDERED column while the INSERTED column contains the updated value. The DELETE_ORDER_LINE trigger uses only the DELETED system table:

```
CREATE TRIGGER DELETE_ORDER_LINE
ON ORDER_LINE
AFTER DELETE
AS

DECLARE @numbord decimal(3,0)
SELECT @numbord = (SELECT NUM_ORDERED FROM DELETED)
UPDATE PART
SET ON_ORDER = ON_ORDER - @numbord
```

Chapter Summary

- There are functions whose results are based on the values in single records. UPPER and LOWER are two examples of functions that act on character data. UPPER displays each lowercase letter in the argument in uppercase. LOWER displays each uppercase letter in the argument in lowercase.
- ROUND and FLOOR are two examples of functions that act on numeric data. ROUND produces its result by rounding the value to the specified number of decimal places. FLOOR produces its result by truncating (removing) everything to the right of the decimal point.
- Use the ADD_MONTHS function in Oracle to add a specific number of months to a date. In Access and in SQL Server, use the DATEADD() function.
- To add a specific number of days to a date, use normal addition. You can subtract one date from another to produce the number of days between two dates.
- To obtain today's date, use the SYSDATE function in Oracle, the GETDATE() function in SQL Server, and the DATE() function in Access.
- To concatenate values in character columns in Oracle, separate the column names with two vertical lines (||). Use the RTRIM function to delete any extra spaces that follow the values. In SQL Server, use the + symbol to concatenate values. In Access, use the ampersand (&) symbol to concatenate values.
- A stored procedure is a query saved in a file that users can execute later.
- To create a stored procedure in PL/SQL or T-SQL, use the CREATE PROCEDURE command.
- Variables in PL/SQL procedures are declared after the word DECLARE. To assign variables the same type as a column in the database, use the %TYPE attribute.
- Use the INTO clause in the SELECT command to place the results of a SELECT command in variables in Oracle.
- You can use INSERT, UPDATE, and DELETE commands in PL/SQL and T-SQL procedures, even when they affect more than one row.
- When a SELECT command is used to retrieve more than one row in PL/SQL or T-SQL, it must define a cursor that will select one row at a time.
- Use the OPEN command to activate a cursor and execute the query in the cursor definition.
- Use the FETCH command to select the next row in PL/SQL and T-SQL.
- Use the CLOSE command to deactivate a cursor. The rows initially retrieved will no longer be available to PL/SQL or T-SQL.
- To use SQL commands in Access, create the command in a string variable. To run the command stored in the string variable, use the DoCmd.RunSQL command.
- To process a collection of rows retrieved by a SELECT command in Access, use a recordset. Create the SQL command in a string variable and use the string variable in the command to open the recordset.
- To move to the next record in a recordset in Access, use the MoveNext command.

- A trigger is an action that occurs automatically in response to an associated database operation, such as an INSERT, UPDATE, or DELETE command. Like a stored procedure, a trigger is stored and compiled on the server. Unlike a stored procedure, which is executed in response to a user request, a trigger is executed in response to a command that causes the associated database operation to occur.

Key Terms

ADD_MONTHS
argument
call
client
client/server system
CLOSE
concatenate
concatenation
cursor
embed
FETCH
FLOOR
LOWER
nonprocedural language
OPEN
PL/SQL
procedural code
procedural language
ROUND
RTRIM
server
stored procedure
SYSDATE
Transact-SQL
trigger
T-SQL
update procedure
UPPER

Review Questions

1. How do you display letters in uppercase in Oracle, Access, and SQL Server? How do you display letters in lowercase in Oracle, Access, and SQL Server?
2. How do you round a number to a specific number of decimal places in Oracle, Access, and SQL Server? How do you remove everything to the right of the decimal place in Oracle and SQL Server?
3. How do you add months to a date in Oracle, Access, and SQL Server? How do you add days to a date? How would you find the number of days between two dates?
4. How do you obtain today's date in Oracle, Access, and SQL Server?
5. How do you concatenate values in character columns in Oracle, Access, and SQL Server?
6. Which function deletes extra spaces at the end of a value?
7. What are stored procedures? What purpose do they serve?
8. In which portion of a PL/SQL procedure do you embed SQL commands?
9. Where do you declare variables in PL/SQL procedures?
10. In PL/SQL, how do you assign variables the same type as a column in the database?
11. How do you place the results of a SELECT command into variables in PL/SQL?

12. Can you use INSERT, UPDATE, or DELETE commands that affect more than one row in PL/SQL procedures?
13. How do you use a SELECT command that retrieves more than one row in a PL/SQL procedure?
14. Which PL/SQL command activates a cursor?
15. Which PL/SQL command selects the next row in a cursor?
16. Which PL/SQL command deactivates a cursor?
17. How do you use SQL commands in Access?
18. How do you process a collection of rows retrieved by a SELECT command in Access?
19. How do you move to the next record in a recordset in Access?
20. What are triggers? What purpose do they serve?
21. What is the purpose of the INSERTED and DELETED tables in SQL Server?

Exercises

Premiere Products

Use the Premiere Products database (see Figure 1-2 in Chapter 1) to complete the following exercises. If directed to do so by your instructor, use the information provided with the Chapter 3 Exercises to print your output.

1. List the part number and description for all parts. The part descriptions should appear in uppercase letters.
2. List the customer number and name for all customers located in the city of Grove. Your query should ignore case. For example, a customer with the city Grove should be included as should customers whose city is GROVE, grove, GrOvE, and so on.
3. List the customer number, name, and balance for all customers. The balance should be rounded to the nearest dollar.
4. Premiere Products is running a promotion that is valid for up to 20 days after an order is placed. List the order number, customer number, customer name, and the promotion date for each order. The promotion date is 20 days after the order was placed.
5. Write PL/SQL or T-SQL procedures to accomplish the following tasks:
   a. Obtain the name and credit limit of the customer whose number currently is stored in I_CUSTOMER_NUM. Place these values in the variables I_CUSTOMER_NAME and I_CREDIT_LIMIT, respectively. Output the contents of I_CUSTOMER_NAME and I_CREDIT_LIMIT.
   b. Obtain the order date, customer number, and name for the order whose number currently is stored in I_ORDER_NUM. Place these values in the variables I_ORDER_DATE, I_CUSTOMER_NUM and I_CUSTOMER_NAME, respectively. Output the contents of I_ORDER_DATE, I_CUSTOMER_NUM, and I_CUSTOMER_NAME.
   c. Add a row to the ORDERS table.

d. Change the date of the order whose number is stored in I_ORDER_NUM to the date currently found in I_ORDER_DATE.

e. Delete the order whose number is stored in I_ORDER_NUM.

6. Write a PL/SQL or T-SQL procedure to retrieve and output the part number, part description, warehouse number, and unit price of every part in the item class stored in I_CLASS.

7. Write Access functions to accomplish the following tasks:

   a. Delete the order whose number is stored in I_ORDER_NUM.

   b. Change the date of the order whose number is stored in I_ORDER_NUM to the date currently found in I_ORDER_DATE.

   c. Retrieve and output the part number, part description, warehouse number, and unit price of every part in the item class stored in I_CLASS.

8. Write a stored procedure in PL/SQL or T-SQL that will change the price of a part with a given part number. How would you use this stored procedure to change the price of part AT94 to $26.95?

9. Write the code for the following triggers in PL/SQL or T-SQL following the style shown in the text:

   a. When adding a customer, add the customer's balance times the sales rep's commission rate to the commission for the corresponding sales rep.

   b. When updating a customer, add the difference between the new balance and the old balance multipled by the sales rep's commission rate to the commission for the corresponding sales rep.

   c. When deleting a customer, subtract the balance multiplied by the sales rep's commission rate from the commission for the corresponding sales rep.

Henry Books

Use the Henry Books database (see Figures 1-4 through 1-7 in Chapter 1) to complete the following exercises. If directed to do so by your instructor, use the information provided with the Chapter 3 Exercises to print your output.

1. List the author number, first name, and last name for all authors. The first name should appear in lowercase letters and the last name should appear in uppercase letters.

2. List the publisher code and name for all publishers located in the city of New York. Your query should ignore case. For example, a customer with the city New York should be included as should customers whose city is NEW YORK, New york, NeW yOrK, and so on.

3. List the book code, title, and price for all books. The price should be rounded to the nearest dollar.

4. Write PL/SQL or T-SQL procedures to accomplish the following tasks:

   a. Obtain the first name and last name of the author whose number currently is stored in I_AUTHOR_NUM. Place these values in the variables I_AUTHOR_FIRST and I_AUTHOR_LAST. Output the contents of I_AUTHOR_NUM, I_AUTHOR_FIRST, and I_AUTHOR_LAST.

b. Obtain the book title, publisher code, and publisher name for every book whose code currently is stored in I_BOOK_CODE. Place these values in the variables I_TITLE, I_PUBLISHER_CODE, and I_PUBLISHER_NAME, respectively. Output the contents of I_TITLE, I_PUBLISHER_CODE, and I_PUBLISHER_NAME.
c. Add a row to the AUTHOR table.
d. Change the last name of the author whose number is stored in I_AUTHOR_NUM to the value currently found in I_AUTHOR_LAST.
e. Delete the author whose number is stored in I_AUTHOR_NUM.

5. Write a PL/SQL or T-SQL procedure to retrieve and output the book code, title, book type, and price for every book whose publisher code is stored in I_PUBLISHER_CODE.
6. Write Access functions to accomplish the following tasks:
   a. Delete the author whose number is stored in I_AUTHOR_NUM.
   b. Change the last name of the author whose number is stored in I_AUTHOR_NUM to the value currently found in I_AUTHOR_LAST.
   c. Retrieve and output the book code, title, book type, and price for every book whose publisher code is stored in I_PUBLISHER_CODE.
7. Write a stored procedure in PL/SQL or T-SQL that will change the price of a book with a given book code. How would you use this stored procedure to change the price of book 0189 to $8.49?
8. Assume the BOOK table contains a column called TOTAL_ON_HAND that represents the total units on hand in all branches for that book. Following the style shown in the text, write the code in PL/SQL or T-SQL for the following triggers:
   a. When inserting a row in the INVENTORY table, add the ON_HAND value to the TOTAL_ON_HAND value for the appropriate book.
   b. When updating a row in the INVENTORY table, add the difference between the new ON_HAND value and the old ON_HAND value to the TOTAL_ON_HAND value for the appropriate book.
   c. When deleting a row in the INVENTORY table, subtract the ON_HAND value from the TOTAL_ON_HAND value for the appropriate book.

Alexamara Marina Group

Use the Alexamara Marina Group database (see Figures 1-8 through 1-12 in Chapter 1) to complete the following exercises. If directed to do so by your instructor, use the information provided with the Chapter 3 Exercises to print your output.

1. List the owner number, first name, and last name for all owners. The first name should appear in uppercase letters and the last name should appear in lowercase letters.
2. List the owner number and last name for all owners located in the city of Bowton. Your query should ignore case. For example, a customer with the city Bowton should be included as should customers whose city is BOWTON, BowTon, BoWtOn, and so on.

3. Alexamara is offering a discount for owners who sign up early for slips for next year. The discount is 1.75 percent of the rental fee. For each slip, list the marina number, slip number, owner number, owner's last name, rental fee, and discount. The discount should be rounded to the nearest dollar.
4. Write PL/SQL or T-SQL procedures to accomplish the following tasks:
   a. Obtain the first name and last name of the owner whose number currently is stored in I_OWNER_NUM. Place these values in the variables I_FIRST_NAME and I_LAST_NAME. Output the contents of I_OWNER_NUM, I_FIRST_NAME, and I_LAST_NAME.
   b. Obtain the marina number, slip number, boat name, owner number, owner first name, and owner last name for the slip whose slip ID is currently stored in I_SLIP_ID. Place these values in the variables I_MARINA_NUM, I_SLIP_NUM, I_BOAT_NAME, I_OWNER_NUM, I_FIRST_NAME, and I_LAST_NAME, respectively. Output the contents of I_SLIP_ID, I_MARINA_NUM, I_SLIP_NUM, I_BOAT_NAME, I_OWNER_NUM, I_FIRST_NAME, and I_LAST_NAME.
   c. Add a row to the OWNER table.
   d. Change the last name of the owner whose number is stored in I_OWNER_NUM to the value currently found in I_LAST_NAME.
   e. Delete the owner whose number is stored in I_OWNER_NUM.
5. Write a PL/SQL or T-SQL procedure to retrieve and output the marina number, slip number, rental fee, boat name, and owner number for every slip whose length is equal to the length stored in I_LENGTH.
6. Write Access functions to accomplish the following tasks:
   a. Delete the owner whose number is stored in I_OWNER_NUM.
   b. Change the last name of the owner whose number is stored in I_OWNER_NUM to the value currently found in I_LAST_NAME.
   c. Retrieve and output the marina number, slip number, rental fee, boat name, and owner number for every slip whose length is equal to the length stored in I_LENGTH.
7. Write a stored procedure in PL/SQL or T-SQL that will change the rental fee of a slip with a given slip ID and marina number. How would you use this stored procedure to change the rental fee for the boat with the slip ID 3 in marina 1 to $3,700?
8. Assume the OWNER table contains a column called TOTAL_RENTAL that represents the total rental fee for all slips rented by that owner. Write the code in PL/SQL or T-SQL for the following triggers following the style shown in the text:
   a. When inserting a row in the MARINA_SLIP table, add the rental fee to the total rental for the appropriate owner.
   b. When updating a row in the MARINA_SLIP table, add the difference between the new rental fee and the old rental fee to the total rental for the appropriate owner.
   c. When deleting a row in the MARINA_SLIP table, subtract the rental fee from the total rental for the appropriate owner.

APPENDIX A

SQL REFERENCE

INTRODUCTION

You can use this appendix to obtain details concerning important components and syntax of the SQL language. Items are arranged alphabetically. Each item contains a description, a reference to where the item is covered in the text, and, when appropriate, both an example and a description of the query results. Some SQL commands also include a description of the clauses associated with them. For each clause, there is a brief description and an indication of whether the clause is required or optional.

ALIASES (PAGES 146–148)

You can specify an alias (alternative name) for each table in a query. You can use the alias in the rest of the command by following the name of the table with a space and the alias name.

The following command creates an alias named R for the REP table and an alias named C for the CUSTOMER table:

```
SELECT R.REP_NUM, R.LAST_NAME, R.FIRST_NAME,
     C.CUSTOMER_NUM, C.CUSTOMER_NAME
FROM REP R, CUSTOMER C
WHERE R.REP_NUM = C.REP_NUM;
```

ALTER TABLE (PAGES 182–183)

Use the ALTER TABLE command to change a table's structure. As shown in Figure A-1, you type the ALTER TABLE command, followed by the table name, and then the alteration to perform.

Clause	Description	Required?
ALTER TABLE *table name*	Indicates name of table to be altered.	Yes
alteration	Indicates type of alteration to be performed.	Yes

FIGURE A-1 ALTER TABLE command

The following command alters the CUSTOMER table by adding a new CUSTOMER_TYPE column:

```
ALTER TABLE CUSTOMER
ADD CUSTOMER_TYPE CHAR(1);
```

The following command changes the CITY column in the CUSTOMER table so that it cannot accept nulls:

```
ALTER TABLE CUSTOMER
MODIFY CITY NOT NULL;
```

Note: In Access, you usually make these changes to a table in Design view rather than using ALTER TABLE.

Note: In SQL Server, you must use the ALTER COLUMN clause and completely define the column as follows:

```
ALTER TABLE CUSTOMER
ALTER COLUMN CITY CHAR(15) NOT NULL
```

COLUMN OR EXPRESSION LIST (SELECT CLAUSE) (PAGES 98–100)

To select columns, use the SELECT clause followed by the list of columns, separated by commas.

The following SELECT clause selects the CUSTOMER_NUM, CUSTOMER_NAME, and BALANCE columns:

```
SELECT CUSTOMER_NUM, CUSTOMER_NAME, BALANCE
```

Use an asterisk in a SELECT clause to select all columns in a table. The following SELECT clause selects all columns:

```
SELECT *
```

Computed Columns (Pages 107–110)

You can use a computation in place of a column by typing the computation. For readability, you can type the computation in parentheses, although it is not necessary to do so.

The following SELECT clause selects the CUSTOMER_NUM and CUSTOMER_NAME columns as well as the results of subtracting the BALANCE column from the CREDIT_LIMIT column:

```
SELECT CUSTOMER_NUM, CUSTOMER_NAME, (CREDIT_LIMIT - BALANCE)
```

The DISTINCT Operator (Pages 117–119)

To avoid selecting duplicate values in a command, use the DISTINCT operator. When you omit the DISTINCT operator from the command and the same value appears on multiple rows in the table, that value will appear on multiple rows in the query results.

The following query selects all customer numbers from the ORDERS table, but lists each customer number only once in the results:

```
SELECT DISTINCT(CUSTOMER_NUM)
FROM ORDERS;
```

Functions (Pages 114–117)

You can use functions in a SELECT clause. The most commonly used functions are AVG (to calculate an average), COUNT (to count the number of rows), MAX (to determine the maximum value), MIN (to determine the minimum value), and SUM (to calculate a total).

The following SELECT clause calculates the average balance:

```
SELECT AVG(BALANCE)
```

COMMIT (PAGES 177–178)

Use the COMMIT command to make permanent any updates made since the last command. If no previous COMMIT command has been executed, the COMMIT command will make all the updates during the current work session permanent immediately. All updates become permanent automatically when you exit SQL. Figure A-2 describes the COMMIT command.

Clause	Description	Required?
COMMIT	Indicates that a COMMIT is to be performed.	Yes

FIGURE A-2 COMMIT command

The following command makes all updates since the most recent COMMIT command permanent:

```
COMMIT;
```

Note: In SQL Server, the following command makes all updates since the most recent COMMIT command:

```
COMMIT TRANSACTION
```

Note: Access does not support the COMMIT command.

CONDITIONS (PAGES 100–106)

A condition is an expression that can be evaluated as either true or false. When you use a condition in a WHERE clause, the results of the query contain those rows for which the condition is true. You can create simple conditions and compound conditions using the BETWEEN, LIKE, IN, EXISTS, ALL, and ANY operators, as described in the following sections.

Simple Conditions (Pages 100–103)

A simple condition has the form column name, comparison operator, and then either another column name or a value. The available comparison operators are = (equal to), < (less than), > (greater than), <= (less than or equal to), >= (greater than or equal to), and < > (not equal to).

The following WHERE clause uses a condition to select rows where the balance is greater than the credit limit:

```
WHERE BALANCE > CREDIT_LIMIT
```

Compound Conditions (Pages 103–106)

Compound conditions are formed by connecting two or more simple conditions using the AND, OR, and NOT operators. When simple conditions are connected by the AND operator, all of the simple conditions must be true in order for the compound condition to be true. When simple conditions are connected by the OR operator, the compound condition will be true whenever any one of the simple conditions is true. Preceding a condition by the NOT operator reverses the truth of the original condition.

The following WHERE clause is true if the warehouse number is equal to 3 *or* the units on hand is greater than 100, *or* both:

```
WHERE (WAREHOUSE = '3') OR (ON_HAND > 100)
```

The following WHERE clause is true if the warehouse number is equal to 3 and the units on hand is greater than 100:

```
WHERE (WAREHOUSE = '3') AND (ON_HAND > 100)
```

The following WHERE clause is true if the warehouse number is not equal to 3:

```
WHERE NOT (WAREHOUSE = '3')
```

BETWEEN Conditions (Pages 106–107)

You can use the BETWEEN operator to determine if a value is within a range of values. The following WHERE clause is true if the balance is between 2,000 and 5,000:

```
WHERE BALANCE BETWEEN 2000 AND 5000
```

LIKE Conditions (Pages 110–111)

LIKE conditions use wildcards to select rows. Use the percent (%) wildcard to represent any collection of characters. The condition LIKE '%Central%' will be true for data consisting of any character or characters, followed by the letters "Central," followed by any other character or characters. Another wildcard symbol is the underscore (_), which represents any individual character. For example, "T_m" represents the letter "T," followed by any single character, followed by the letter "m," and would be true for a collection of characters such as Tim, Tom, or T3m.

The following WHERE clause is true if the value in the STREET column is Central, Centralia, or any other value that contains "Central":

```
WHERE STREET LIKE '%Central%'
```

Note: Access uses different wildcard symbols. The symbol for any collection of characters is the asterisk (*). The symbol for an individual character is the question mark (?).

IN Conditions (Pages 111–112, 140–141)

You can use IN to determine whether a value is in some specific collection of values. The following WHERE clause is true if the credit limit is 5,000, 10,000, or 15,000:

```
WHERE CREDIT_LIMIT IN (5000, 10000, 15000)
```

The following WHERE clause is true if the part number is in the collection of part numbers associated with order number 21610:

```
WHERE PART_NUM IN
(SELECT PART_NUM
FROM ORDER_LINE
WHERE ORDER_NUM = '21610')
```

EXISTS Conditions (Pages 141–142)

You can use EXISTS to determine whether the results of a subquery contain at least one row. The following WHERE clause is true if the results of the subquery contain at least one row, that is, there is at least one order line with the desired order number and on which the part number is DR93:

```
WHERE EXISTS
(SELECT *
FROM ORDER_LINE
WHERE ORDERS.ORDER_NUM = ORDER_LINE.ORDER_NUM
AND PART_NUM = 'DR93')
```

ALL and ANY (Pages 157–160)

You can use ALL or ANY with subqueries. If you precede the subquery by ALL, the condition is true only if it is satisfied for all values produced by the subquery. If you precede the subquery by ANY, the condition is true if it is satisfied for any value (one or more) produced by the subquery.

The following WHERE clause is true if the balance is greater than every balance contained in the results of the subquery:

```
WHERE BALANCE > ALL
(SELECT BALANCE
FROM CUSTOMER
WHERE REP_NUM = '65')
```

The following WHERE clause is true if the balance is greater than at least one balance contained in the results of the subquery:

```
WHERE BALANCE > ANY
(SELECT BALANCE
FROM CUSTOMER
WHERE REP_NUM = '65')
```

CREATE INDEX (PAGES 215–217)

Use the CREATE INDEX command to create an index for a table. Figure A-3 describes the CREATE INDEX command.

Clause	Description	Required?
CREATE INDEX *index name*	Indicates the name of the index.	Yes
ON *table name*	Indicates the table for which the index is to be created.	Yes
column list	Indicates the column or columns on which the index is to be based.	Yes

FIGURE A-3 CREATE INDEX command

The following CREATE INDEX command creates an index named REPNAME for the REP table on the combination of the LAST_NAME and FIRST_NAME columns:

```
CREATE INDEX REPNAME ON REP(LAST_NAME, FIRST_NAME);
```

CREATE TABLE (PAGES 66–69)

Use the CREATE TABLE command to define the structure of a new table. Figure A-4 describes the CREATE TABLE command.

Clause	Description	Required?
CREATE TABLE *table name*	Indicates the name of the table to be created.	Yes
(*column and data type list*)	Indicates the columns that comprise the table along with their corresponding data types (see Data Types section).	Yes

FIGURE A-4 CREATE TABLE command

The following CREATE TABLE command creates the REP table and its associated columns and data types. REP_NUM is the table's primary key.

```
CREATE TABLE REP
(REP_NUM CHAR(2) PRIMARY KEY,
LAST_NAME CHAR(15),
FIRST_NAME CHAR(15),
STREET CHAR(15),
CITY CHAR(15),
STATE CHAR(2),
ZIP CHAR(5),
COMMISSION DECIMAL(7,2),
RATE DECIMAL(3,2) );
```

Note: Access does not support the DECIMAL data type. Use the CURRENCY data type for fields that will contain currency values; use the NUMBER data type for all other numeric fields. In Access, use the following command to create the REP table:

```
CREATE TABLE REP
(REP_NUM CHAR(2) PRIMARY KEY,
LAST_NAME CHAR(15),
FIRST_NAME CHAR(15),
STREET CHAR(15),
CITY CHAR(15),
STATE CHAR(2),
ZIP CHAR(5),
COMMISSION CURRENCY,
RATE NUMBER );
```

CREATE VIEW (PAGES 196–197)

Use the CREATE VIEW command to create a view. Figure A-5 describes the CREATE VIEW command.

Clause	Description	Required?
CREATE VIEW *view name* AS	Indicates the name of the view to be created.	Yes
query	Indicates the defining query for the view.	Yes

FIGURE A-5 CREATE VIEW command

The following CREATE VIEW command creates a view named HOUSEWARES, which consists of the part number, part description, units on hand, and unit price for all rows in the PART table on which the item class is HW:

```
CREATE VIEW HOUSEWARES AS
SELECT PART_NUM, PART_DESCRIPTION, ON_HAND, PRICE
FROM PART
WHERE CLASS = 'HW';
```

DATA TYPES (PAGE 71)

Figure A-6 describes the data types that you can use in a CREATE TABLE command.

Data Type	Description
CHAR(*n*)	Stores a character string *n* characters long. You use the CHAR type for columns that contain letters and special characters and for columns containing numbers that will not be used in any calculations. Because neither sales rep numbers nor customer numbers will be used in any calculations, for example, the REP_NUM and CUSTOMER_NUM columns are both assigned theCHAR data type.
VARCHAR(*n*)	An alternative to CHAR that stores a character string up to *n* characters long. Unlike CHAR, only the actual character string is stored. If a character string 20 characters long is stored in a CHAR(30) column, for example, it will occupy 30 characters (20 characters plus 10 blank spaces). If it is stored in a VARCHAR(30) column, it will only occupy 20 spaces. In general, tables that use VARCHAR instead of CHAR occupy less space, but the DBMS does not process them as rapidly during queries and updates. However, both are legitimate choices. This text uses CHAR, but VARCHAR would work equally well.
DATE	Stores date data. The specific format in which dates are stored varies from one SQL implementation to another. In Oracle, dates are enclosed in single quotation marks and have the form DD-MON-YYYY (for example,'15-OCT-2010' is October 15, 2010). In Access, dates are enclosed in number signs and are entered using the format MM/DD/YYYY (for example, #10/15/2010# is October 15, 2010). In SQL Server, use the DATETIME data type to store dates.
DECIMAL(*p*,*q*)	Stores a decimal number *p* digits long with *q* of these digits being decimal places to the right of the decimal point. For example, the data type DECIMAL(5,2) represents a number with three places to the left and two places to the right of the decimal (for example, 100.00). You can use the contents of DECIMAL columns in calculations. You also can use the NUMBER(*p*,*q*) data type in both Oracle and SQL Server to store a decimal number. Access does not support the DECIMAL data type; use the CURRENCY or NUMBER data type instead.
INT	Stores integers, which are numbers without a decimal part. The valid range is -2147483648 to 2147483647. You can use the contents of INT columns in calculations. If you follow the word INT with AUTO_INCREMENT, you create a column for which SQL will automatically generate a new sequence number each time you add a new row. This would be the appropriate choice, for example, when you want the DBMS to generate a value for a primary key.
SMALLINT	Stores integers, but uses less space than the INT data type. The valid range is -32768 to 32767. SMALLINT is a better choice than INT when you are certain that the column will store numbers within the indicated range. You can use the contents of SMALLINT columns in calculations.

FIGURE A-6 Data types

DELETE ROWS (PAGES 79, 178–180)

Use the DELETE command to delete one or more rows from a table. Figure A-7 describes the DELETE command.

Clause	Description	Required?
DELETE *table name*	Indicates the table from which the row or rows are to be deleted.	Yes
WHERE *condition*	Indicates a condition. Those rows for which the condition is true will be retrieved and deleted.	No (If you omit the WHERE clause, all rows will be deleted.)

FIGURE A-7 DELETE command

The following DELETE command deletes any row from the LEVEL1_CUSTOMER table on which the customer number is 842:

```
DELETE LEVEL1_CUSTOMER
WHERE CUSTOMER_NUM = '842';
```

DESCRIBE (PAGES 87–88)

In Oracle, you can use the DESCRIBE command to list all the columns in a table and their properties. The following command describes the REP table:

```
DESCRIBE REP;
```

Note: In Access, use the Documenter to describe the tables and other objects in a database.

Note: In SQL Server, execute the sp_columns command to list all the columns in a table. The following command will list all the columns in the REP table:

```
Exec sp_columns REP
```

DROP INDEX (PAGE 217)

Use the DROP INDEX command to delete an index, as shown in Figure A-8.

Clause	Description	Required?
DROP INDEX *index name*	Indicates the name of the index to be dropped.	Yes

FIGURE A-8 DROP INDEX command

The following DROP INDEX command deletes the index named CREDNAME:

```
DROP INDEX CREDNAME;
```

Note: In SQL Server, you must qualify the index name as follows:

```
DROP INDEX CUSTOMER.CREDNAME
```

DROP TABLE (PAGES 70, 189)

Use the DROP TABLE command to delete a table, as shown in Figure A-9.

Clause	Description	Required?
DROP TABLE *table name*	Indicates name of the table to be dropped.	Yes

FIGURE A-9 DROP TABLE command

The following DROP TABLE command deletes the table named LEVEL1_CUSTOMER:

```
DROP TABLE LEVEL1_CUSTOMER;
```

DROP VIEW (PAGES 208–209)

Use the DROP VIEW command to delete a view, as shown in Figure A-10.

Clause	Description	Required?
DROP VIEW *view name*	Indicates the name of the view to be dropped.	Yes

FIGURE A-10 DROP VIEW command

The following DROP VIEW command deletes the view named HSEWRES:

```
DROP VIEW HSEWRES;
```

GRANT (PAGES 209–212)

Use the GRANT command to grant privileges to a user. Figure A-11 describes the GRANT command.

Clause	Description	Required?
GRANT *privilege*	Indicates the type of privilege(s) to be granted.	Yes
ON *database object*	Indicates the database object(s) to which the privilege(s) pertain.	Yes
TO *user name*	Indicates the user(s) to whom the privilege(s) are to be granted. To grant the privilege(s) to all users, use the TO PUBLIC clause.	Yes

FIGURE A-11 GRANT command

The following GRANT command grants the user named Johnson the privilege of selecting rows from the REP table:

```
GRANT SELECT
ON REP
TO Johnson;
```

INSERT INTO (QUERY) (PAGES 172–173)

Use the INSERT INTO command with a query to insert the rows retrieved by a query into a table. As shown in Figure A-12, you must indicate the name of the table into which the row(s) will be inserted and the query whose results will be inserted into the named table.

Clause	Description	Required?
INSERT INTO *table name*	Indicates the name of the table into which the row(s) will be inserted.	Yes
query	Indicates the query whose results will be inserted into the table.	Yes

FIGURE A-12 INSERT INTO (query) command

The following INSERT INTO command inserts rows selected by a query into the LEVEL1_CUSTOMER table:

```
INSERT INTO LEVEL1_CUSTOMER
SELECT CUSTOMER_NUM, CUSTOMER_NAME, BALANCE,
     CREDIT_LIMIT, REP_NUM
FROM CUSTOMER
WHERE CREDIT_LIMIT = 7500;
```

INSERT INTO (VALUES) (PAGES 72–75)

Use the INSERT INTO command and the VALUES clause to insert a row into a table by specifying the values for each of the columns. As shown in Figure A-13, you must indicate the table into which to insert the values, and then list the values to insert in parentheses.

Clause	Description	Required?
INSERT INTO *table name*	Indicates the name of the table into which the row will be inserted.	Yes
VALUES (*values list*)	Indicates the values for each of the columns on the new row.	Yes

FIGURE A-13 INSERT INTO (values) command

The following INSERT INTO command inserts the values shown in parentheses as a new row in the REP table:

```
INSERT INTO REP
VALUES
('20','Kaiser','Valerie','624 Randall','Grove','FL','33321',20542.50,0.05);
```

INTEGRITY (PAGES 221–225)

You can use the ALTER TABLE command with an appropriate ADD CHECK, ADD PRIMARY KEY, or ADD FOREIGN KEY clause to specify integrity. Figure A-14 describes the ALTER TABLE command for specifying integrity.

Clause	Description	Required?
ALTER TABLE *table name*	Indicates the table for which integrity is being specified.	Yes
integrity clause	ADD CHECK, ADD PRIMARY KEY, or ADD FOREIGN KEY	Yes

FIGURE A-14 Integrity options

The following ALTER TABLE command changes the PART table so that the only legal values for the CLASS column are AP, HW, and SG:

```
ALTER TABLE PART
ADD CHECK (CLASS IN ('AP','HW','SG') );
```

The following ALTER TABLE command changes the REP table so that the REP_NUM column is the table's primary key:

```
ALTER TABLE REP
ADD PRIMARY KEY(REP_NUM);
```

The following ALTER TABLE command changes the CUSTOMER table so that the REP_NUM column in the CUSTOMER table is a foreign key referencing the primary key of the REP table:

```
ALTER TABLE CUSTOMER
ADD FOREIGN KEY(REP_NUM) REFERENCES REP;
```

REVOKE (PAGES 209–212)

Use the REVOKE command to revoke privileges from a user. Figure A-15 describes the REVOKE command.

Clause	Description	Required?
REVOKE *privilege*	Indicates the type of privilege(s) to be revoked.	Yes
ON *database object*	Indicates the database object(s) to which the privilege pertains.	Yes
FROM *user name*	Indicates the user name(s) from whom the privilege(s) are to be revoked.	Yes

FIGURE A-15 REVOKE command

The following REVOKE command revokes the SELECT privilege for the REP table from the user named Johnson:

```
REVOKE SELECT
ON REP
FROM Johnson;
```

ROLLBACK (PAGES 177–178)

Use the ROLLBACK command to reverse (undo) all updates since the execution of the previous COMMIT command. If no COMMIT command has been executed, the command will undo all changes made during the current work session. Figure A-16 describes the ROLLBACK command.

Clause	Description	Required?
ROLLBACK	Indicates that a rollback is to be performed.	Yes

FIGURE A-16 ROLLBACK command

The following command reverses all updates made since the time of the last COMMIT command:

```
ROLLBACK;
```

Note: In SQL Server, the following command reverses all updates made since the time of the last COMMIT command:

```
ROLLBACK TRANSACTION
```

Note: Access does not support the ROLLBACK command.

SELECT (PAGES 75–78, 98–127)

Use the SELECT command to retrieve data from a table or from multiple tables. Figure A-17 describes the SELECT command.

Clause	Description	Required?
SELECT *column or expression list*	Indicates the column(s) and/or expression(s) to be retrieved.	Yes
FROM *table list*	Indicates the table(s) required for the query.	Yes
WHERE *condition*	Indicates one or more conditions. Only the rows for which the condition(s) are true will be retrieved.	No (If you omit the WHERE clause, all rows will be retrieved.)
GROUP BY *column list*	Indicates column(s) on which rows are to be grouped.	No (If you omit the GROUP BY clause, no grouping will occur.)
HAVING *condition involving groups*	Indicates a condition for groups. Only groups for which the condition is true will be included in query results. Use the HAVING clause only if the query output is grouped.	No (If you omit the HAVING clause, all groups will be included.)
ORDER BY *column or expression list*	Indicates column(s) on which the query output is to be sorted.	No (If you omit the ORDER BY clause, no sorting will occur.)

FIGURE A-17 SELECT command

The following SELECT command joins the ORDERS and ORDER_LINE tables. The command selects the customer number, order number, order date, and the sum of the product of the number ordered and unit price, renamed as ORDER_TOTAL. Records are grouped by order number, customer number, and order date. Only groups on which the order total is greater than 1,000 are included. Groups are ordered by order number.

```
SELECT CUSTOMER_NUM, ORDERS.ORDER_NUM, ORDER_DATE,
     SUM(NUM_ORDERED * QUOTED_PRICE) AS ORDER_TOTAL
FROM ORDERS, ORDER_LINE
WHERE ORDERS.ORDER_NUM = ORDER_LINE.ORDER_NUM
GROUP BY ORDERS.ORDER_NUM, CUSTOMER_NUM, ORDER_DATE
HAVING SUM(NUM_ORDERED * QUOTED_PRICE) > 1000
ORDER BY ORDERS.ORDER_NUM;
```

SUBQUERIES (PAGES 120–123, 142–144)

You can use one query within another. The inner query is called a subquery and it is evaluated first. The outer query is evaluated next. The following command contains a subquery that produces a list of part numbers included in order number 21610:

```
SELECT PART_DESCRIPTION
FROM PART
WHERE PART_NUM IN
(SELECT PART_NUM
FROM ORDER_LINE
WHERE ORDER_NUM = '21610');
```

UNION, INTERSECT, AND MINUS (PAGES 152–157)

Connecting two SELECT commands with the UNION operator produces all the rows that would be in the results of the first query, the second query, or both queries. Connecting two SELECT commands with the INTERSECT operator produces all the rows that would be in the results of both queries. Connecting two SELECT commands with the MINUS operator produces all the rows that would be in the results of the first query, but *not* in the results of the second query. Figure A-18 describes the UNION, INTERSECT, and MINUS operators.

Operator	Description
UNION	Produces all the rows that would be in the results of the first query, the second query, or both queries.
INTERSECT	Produces all the rows that would be in the results of both queries.
MINUS	Produces all the rows that would be in the results of the first query but not in the results of the second query.

FIGURE A-18 UNION, INTERSECT, and MINUS operators

Note: Access and SQL Server support the UNION operation. SQL Server supports the INTERSECT command while Access does not. Neither SQL Server nor Microsoft Access support the MINUS operator.

The following query displays the customer number and customer name of all customers that are represented by sales rep 65, *or* that have orders, *or* both:

```
SELECT CUSTOMER_NUM, CUSTOMER_NAME
FROM CUSTOMER
WHERE REP_NUM = '65'
UNION
SELECT CUSTOMER.CUSTOMER_NUM, CUSTOMER_NAME
FROM CUSTOMER, ORDERS
WHERE CUSTOMER.CUSTOMER_NUM = ORDERS.CUSTOMER_NUM;
```

The following query displays the customer number and customer name of all customers that are represented by sales rep 65 *and* that have orders:

```
SELECT CUSTOMER_NUM, CUSTOMER_NAME
FROM CUSTOMER
WHERE REP_NUM = '65'
INTERSECT
SELECT CUSTOMER.CUSTOMER_NUM, CUSTOMER_NAME
FROM CUSTOMER, ORDERS
WHERE CUSTOMER.CUSTOMER_NUM = ORDERS.CUSTOMER_NUM;
```

The following query displays the customer number and customer name of all customers that are represented by sales rep 65 but that do *not* have orders:

```
SELECT CUSTOMER_NUM, CUSTOMER_NAME
FROM CUSTOMER
WHERE REP_NUM = '65'
MINUS
SELECT CUSTOMER.CUSTOMER_NUM, CUSTOMER_NAME
FROM CUSTOMER, ORDERS
WHERE CUSTOMER.CUSTOMER_NUM = ORDERS.CUSTOMER_NUM;
```

UPDATE (PAGES 78–80, 173–175)

Use the UPDATE command to change the contents of one or more rows in a table. Figure A-19 describes the UPDATE command.

Clause	Description	Required?
UPDATE *table name*	Indicates the table whose contents will be changed.	Yes
SET *column = expression*	Indicates the column to be changed, along with an expression that provides the new value.	Yes
WHERE *condition*	Indicates a condition. The change will occur only on those rows for which the condition is true.	No (If you omit the WHERE clause, all rows will be updated.)

FIGURE A-19 UPDATE command

The following UPDATE command changes the customer name on the row in LEVEL1_CUSTOMER on which the customer number is 842 to All Season Sport:

```
UPDATE LEVEL1_CUSTOMER
SET CUSTOMER_NAME = 'All Season Sport'
WHERE CUSTOMER_NUM = '842';
```

HOW DO I REFERENCE

This appendix answers frequently asked questions about how to accomplish a variety of tasks using SQL.

Use the second column to locate the correct section in Appendix A that answers your question.

How Do I	Review the Named Section(s) in Appendix A
Add columns to an existing table?	ALTER TABLE
Add rows?	INSERT INTO (Values)
Calculate a statistic (sum, average, maximum, minimum, or count)?	1. SELECT 2. Column or Expression List (SELECT Clause) (Use the appropriate function in the query.)
Change rows?	UPDATE
Create a data type for a column?	1. Data Types 2. CREATE TABLE
Create a table?	CREATE TABLE
Create a view?	CREATE VIEW
Create an index?	CREATE INDEX
Describe a table's layout?	DESCRIBE
Delete a table?	DROP TABLE
Delete a view?	DROP VIEW
Delete an index?	DROP INDEX
Delete rows?	DELETE Rows
Drop a table?	DROP TABLE
Drop a view?	DROP VIEW
Drop an index?	DROP INDEX
Grant a privilege?	GRANT
Group data in a query?	SELECT (Use a GROUP BY clause.)
Insert rows?	INSERT INTO (Values)
Insert rows using a query?	INSERT INTO (Query)
Join tables?	Conditions (Include a WHERE clause to relate the tables.)
Make updates permanent?	COMMIT
Order query results?	SELECT (Use the ORDER BY clause.)
Prohibit nulls?	1. CREATE TABLE 2. ALTER TABLE (Include the NOT NULL clause in a CREATE TABLE or ALTER TABLE command.)
Remove a privilege?	REVOKE
Remove rows?	DELETE Rows
Retrieve all columns?	1. SELECT 2. Column or Expression List (SELECT Clause) (Type *in the SELECT clause.)

FIGURE B-1 How Do I reference

How Do I	Review the Named Section(s) in Appendix A
Retrieve all rows?	SELECT (Omit the WHERE clause.)
Retrieve only certain columns?	1. SELECT 2. Column or Expression List (SELECT Clause) (Type the list of columns in the SELECT clause.)
Revoke a privilege?	REVOKE
Select all columns?	1. SELECT 2. Column or Expression List (SELECT Clause) (Type *in the SELECT clause.)
Select all rows?	SELECT (Omit the WHERE clause.)
Select only certain columns?	1. SELECT 2. Column or Expression List (SELECT Clause) (Type the list of columns in the SELECT clause.)
Select only certain rows?	1. SELECT 2. Conditions (Use a WHERE clause.)
Sort query results?	SELECT (Use an ORDER BY clause.)
Specify a foreign key?	Integrity (Use the ADD FOREIGN KEY clause in an ALTER TABLE command.)
Specify a primary key?	Integrity (Use the ADD PRIMARY KEY clause in an ALTER TABLE command.)
Specify a privilege?	GRANT
Specify integrity?	Integrity (Use an ADD CHECK, ADD PRIMARY KEY, and/or ADD FOREIGN KEY clause in an ALTER TABLE command.)
Specify legal values?	Integrity (Use an ADD CHECK clause in an ALTER TABLE command.)
Undo updates?	ROLLBACK
Update rows?	UPDATE
Use a calculated field?	1. SELECT 2. Column or Expression List (SELECT Clause) (Enter a calculation in the query.)
Use a compound condition?	1. SELECT 2. Conditions (Use simple conditions connected by AND, OR, or NOT in a WHERE clause.)
Use a compound condition in a query?	Conditions

FIGURE B-1 How Do I reference (continued)

How Do I	Review the Named Section(s) in Appendix A
Use a condition in a query?	1. SELECT 2. Conditions (Use a WHERE clause.)
Use a subquery?	Subqueries
Use a wildcard?	1. SELECT 2. Conditions (Use LIKE and a wildcard in a WHERE clause.)
Use an alias?	Aliases (Enter an alias after the name of each table in the FROM clause.)
Use set operations (union, intersection, difference)?	UNION, INTERSECT, and MINUS (Connect two SELECT commands with UNION, INTERSECT, or MINUS.)

FIGURE B-1 How Do I reference (continued)

ANSWERS TO ODD-NUMBERED REVIEW QUESTIONS

CHAPTER 1—INTRODUCTION TO PREMIERE PRODUCTS, HENRY BOOKS, AND ALEXAMARA MARINA GROUP

Due to the nature of the material in Chapter 1, there are no Review Questions.

CHAPTER 2—DATABASE DESIGN FUNDAMENTALS

1. An entity is a person, place, thing, or event.
3. A relationship is an association between tables (entities). A one-to-many relationship between two tables is a relationship in which each row in the first table can be associated with many rows in the second table, but each row in the second table is associated with only one row in the first table.
5. A relation is a two-dimensional table in which the entries in the table are single-valued (each location in the table contains a single entry), each column has a distinct name (or attribute name), all values in a column match this name, the order of the rows and columns is immaterial, and each row contains unique values.
7. For each table, you write the name of the table and then within parentheses list all of the columns in the table. Underline the primary keys.

```
BRANCH (BRANCH_NUM, BRANCH_NAME, BRANCH_LOCATION,
       NUM_EMPLOYEES)
PUBLISHER (PUBLISHER_CODE, PUBLISHER_NAME, CITY)
AUTHOR (AUTHOR_NUM, AUTHOR_LAST, AUTHOR_FIRST)
BOOK (BOOK_CODE, TITLE, PUBLISHER_CODE, TYPE, PRICE,
       PAPERBACK)
WROTE (BOOK_CODE, AUTHOR_NUM, SEQUENCE)
INVENTORY (BOOK_CODE, BRANCH_NUM, ON_HAND)
```

9. A column (attribute), B, is functionally dependent on another column (or a collection of columns), A, if at any point in time a value for A determines a single value for B.

11. Functional dependencies:

```
DEPARTMENT_NUM → DEPARTMENT_NAME
ADVISOR_NUM → ADVISOR_LAST_NAME, ADVISOR_FIRST_NAME,
        DEPARTMENT_NUM
COURSE_CODE → DESCRIPTION
STUDENT_NUM → STUDENT_LAST_NAME, STUDENT_FIRST_NAME,
        ADVISOR_NUM
STUDENT_NUM, COURSE_CODE → GRADE
```

Relations:

```
DEPARTMENT (DEPARTMENT_NUM, DEPARTMENT_NAME)
ADVISOR (ADVISOR_NUM, ADVISOR_LAST_NAME, ADVISOR_FIRST_NAME,
        DEPARTMENT_NUM)
COURSE (COURSE_CODE, DESCRIPTION)
STUDENT (STUDENT_NUM, STUDENT_LAST_NAME, STUDENT_FIRST_NAME,
        ADVISOR_NUM)
STUDENT_COURSE (STUDENT_NUM, COURSE_CODE, GRADE)
```

Entity-relationship diagram: (*Note:* Your rectangles can be in different positions as long as they are connected by the same arrows.)

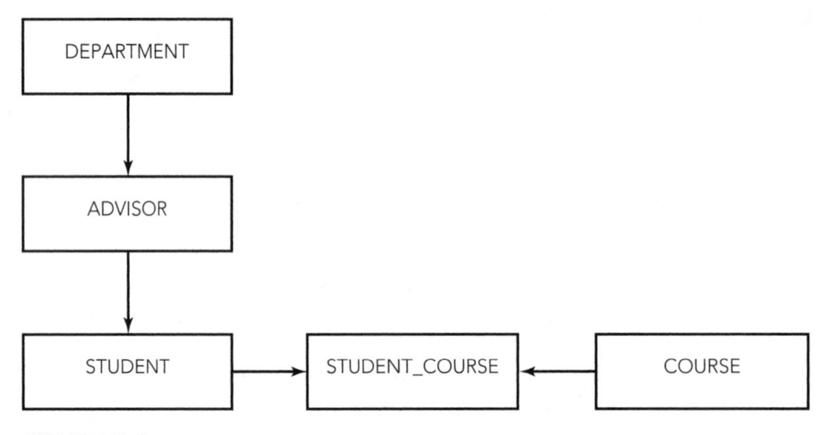

FIGURE C-1

13. A table (relation) is in second normal form when it is in first normal form and no nonkey column is dependent on only a portion of the primary key. When a table is not in second normal form, the table contains redundancy, which leads to a variety of update anomalies. A change in a value can require not just one change, but several. There is the possibility of inconsistent data. Adding additional data to the database might not be possible without creating artificial values for part of the key. Finally, deletions of certain items can result in inadvertently deleting crucial information from the database.

15.

```
STUDENT (STUDENT_NUM, STUDENT_LAST_NAME, STUDENT_FIRST_NAME,
        ADVISOR_NUM)
ADVISOR (ADVISOR_NUM, ADVISOR_LAST_NAME, ADVISOR_FIRST_NAME)
COURSE (COURSE_CODE, DESCRIPTION)
STUDENT_COURSE (STUDENT_NUM, COURSE_CODE, GRADE)
```

CHAPTER 3—CREATING TABLES

1. Use the CREATE TABLE command to create a table by typing the table name and then listing within a single set of parentheses the columns in the table.
3. CHAR, VARCHAR, DATE, DECIMAL, INT, SMALLINT
5. Answers will vary. Answers should mention that the difference between CHAR and VARCHAR is that CHAR is fixed length, while VARCHAR is variable length. This means that CHAR is always the same size and takes up the same amount of bytes, while VARCHAR varies. VARCHAR is a good choice when you are storing email addresses and comments that can vary in size.
7. Use the INSERT command.
9. Use the UPDATE command.
11. Use the DESCRIBE command.

CHAPTER 4—SINGLE-TABLE QUERIES

1. The basic form of the SELECT command is SELECT-FROM-WHERE. Specify the columns to be listed after the word SELECT (or type * to select all columns), and then specify the table name that contains these columns after the word FROM. Optionally, you can include condition(s) after the word WHERE.
3. You can form a compound condition by combining simple conditions and using the operators AND, OR, or NOT.
5. Use arithmetic operators and write the computation in place of a column name. You can assign a name to the computation by following the computation with the word AS and then the desired name.
7. In Oracle, the percent (%) wildcard represents any collection of characters. The underscore (_) wildcard represents any single character.
9. Use an ORDER BY clause.
11. To sort data in descending order, follow the sort key with the DESC operator.
13. To avoid duplicates, precede the column name with the DISTINCT operator.
15. Use a GROUP BY clause.
17. Use the IS NULL operator in the WHERE clause.

CHAPTER 5—MULTIPLE-TABLE QUERIES

1. Indicate in the SELECT clause all columns to display, list in the FROM clause all tables to join, and then include in the WHERE clause any conditions requiring values in matching columns to be equal.
3. IN and EXISTS
5. An alias is an alternate name for a table. To specify an alias in SQL, follow the name of the table with the name of the alias. You use the alias just like a table name throughout the SQL command.
7. Use the UNION, INTERSECT, and MINUS operators to create a union, intersection, and difference of two tables. To perform any of these operations, the tables must be union compatible.

9. When the ALL operator precedes a subquery, the condition is true only if it satisfies all values produced by the subquery.
11. In an inner join, only matching rows from both tables are included. You can use the INNER JOIN clause to perform an inner join.
13. In a right outer join, all rows from the table on the right will be included regardless of whether they match rows from the table on the left. Rows from the table on the left will be included only if they match. You can use the RIGHT JOIN clause to perform a right outer join.
15. Answers will vary. Answers should note that an equi-join is similar to an inner join except that both matching columns appear in the results. A natural join is the same as the inner join discussed in Chapter 5. A cross join is the same as a Cartesian product.

CHAPTER 6—UPDATING DATA

1. CREATE TABLE
3. Use the INSERT command with a SELECT clause.
5. DELETE
7. In Oracle, use the ROLLBACK command. In SQL Server, use the ROLLBACK TRANSACTION command. Any updates made since the most recent COMMIT command (or COMMIT TRANSACTION command in SQL Server) are reversed.
9. The clause is SET followed by the column name, followed by an equals sign (=) and the word NULL.
11. In Oracle, use the ALTER TABLE command with a MODIFY clause. In SQL Server, use the ALTER TABLE command with an ALTER COLUMN clause.
13. Use a make-table query to create a table from another table. The equivalent SQL commands to the Access make-table query are CREATE TABLE, SELECT, and INSERT.

CHAPTER 7—DATABASE ADMINISTRATION

1. A view contains data that is derived from existing base tables when users attempt to access the view.
3. A defining query is the portion of the CREATE VIEW command that describes the data to include in a view.
5. Views provide data independence, allow database access control, and simplify the database structure for users.
7. DROP VIEW
9. REVOKE
11. Use the CREATE INDEX command to create an index. Use the CREATE UNIQUE INDEX command to create a unique index. A unique index allows only unique values in the column (or columns) on which the index is created.
13. The DBMS
15. Answers will vary. Answers should note that a data dictionary is a catalog that stores data about the entities, attributes, relationships, programs, and other

objects in a database. Some items found in a data dictionary include synonyms for attributes, detailed descriptions of each table and attribute in the database, referential integrity constraints, and database schema definitions.

17. The DBMS updates the system catalog automatically when users make change to the database, such as creating, altering, or dropping tables or creating or dropping indexes.
19. Use the CHECK clause of the ALTER TABLE command.
21. Use the ADD FOREIGN KEY clause of the ALTER TABLE command.

CHAPTER 8—SQL FUNCTIONS AND PROCEDURES

1. Use the UPPER function to display letters in uppercase in Oracle and SQL Server. In Access, use the UCASE() function. Use the LOWER function to display letters in lowercase in Oracle and SQL Server. In Access, use the LCASE() function.
3. To add months to a date, use the ADD_MONTHS function (Oracle), or the DATEADD() function (Access and SQL Server). To add days to a date, add the desired number of days to a date. To find the number of days between two dates, subtract the earlier date from the later date.
5. In Oracle, separate the column names with two vertical lines (||) in the SELECT clause. In SQL Server, separate the column names with the + symbol. In Access, separate the column names with the & symbol.
7. A stored procedure is a file that is stored on a server and contains commands that can be used repeatedly. Stored procedures eliminate the need for users to retype a query each time it is needed.
9. In PL/SQL procedures, you declare variables first before any procedural code.
11. Use the INTO clause to place the results of a SELECT statement in variables.
13. When retrieving multiple rows with a SELECT statement, use a cursor.
15. FETCH
17. To use SQL commands in Access, create the command in a string variable. To run the command stored in the string variable, use the DoCmd.RunSQL command.
19. To move to the next record in an Access recordset, use the MoveNext command.
21. The INSERTED and DELETED tables are temporary system tables created by SQL Server. The INSERTED table contains the most recent (updated) values in a record and the DELETED table contains the previous (before update) value.

INDEX

B

C

D

E

F

G

H

I

J

K

L

M

N

O

P

Q

R

S

T

U

V

W